Q175 FRA

SELECTIVITY AND DISCORD

Selectivity AND Discord

Two Problems of Experiment

ALLAN FRANKLIN

University of Pittsburgh Press

Published by the University of Pittsburgh Press, Pittsburgh, Pa. 15261

Copyright © 2002, University of Pittsburgh Press

Manufactured in the United States of America

Printed on acid-free paper

10   9   8   7   6   5   4   3   2   1

Library of Congress Cataloging-in-Publication Data

Franklin, Allan, 1938–

   Selectivity and discord : two problems of experiment / Allan Franklin

      p.   cm.

   Includes bibliographical references and index.

   ISBN 0-8229-4191-0 (cloth : alk. paper)

   1. Science—Philosophy.   2. Physics—Experiments.   3. Physics—History.

I. Title

Q175.F795 2002

   501—dc21

2002011054

To Cyndi Betts, my wife and best friend,
without whom this work could not have been done

CONTENTS

ACKNOWLEDGMENTS

Some of the work for this book was carried out while I was a resident fellow at the Dibner Institute for the History of Science and Technology. I am grateful to Jed Buchwald, the director; Evelyn Simha, the executive director; and their staff, Carla Chrisfield, Rita Dempsey, and Trudy Kontoff, for providing both support and an atmosphere in which it is almost impossible not to get work done. I have benefited enormously from discussions with Rudolf Ganz, one of the leading experimenters on possible low-mass electron-positron states, who also provided a copy of his dissertation. I have also had valuable discussions with George Smith; Roger Stuewer; Noel Swerdlow; my colleagues in the Department of Physics at the University of Colorado, Tony Barker, John Cumalat, and Patricia Rankin; and my colleagues in the Department of History and Philosophy of Science at the University of Melbourne, Maureen Christie, Rod Home, Howard Sankey, and Neil Thomason; and Manuel Thomaz, who joined me as a visitor in the department. Much of the material in this book originally appeared in my papers "The Resolution of Discordant Results," *Perspectives on Science* (1995) 3: 346–420, and "Selectivity and the Production of Experimental Results," *Archive for the History of Exact Sciences* (1998) 53: 399–485. The material in Chapter 8 appeared in "William Wilson and the Absorption of Beta Rays," *Physics in Perspective* (2002) 4: 40–70.

SELECTIVITY AND DISCORD

Introduction

Physics and, I believe, all of science is a reasonable enterprise based on experimental evidence, criticism, and rational discussion. It provides us with knowledge of the physical world, and it is experiment that provides the evidence that grounds this knowledge. As the late Richard Feynman, one of the leading theoretical physicists of the 20th century, wrote, "The principle of science, the definition, almost, is the following: *The test of all knowledge is experiment.* Experiment is the *sole judge* of scientific 'truth'" (Feynman et al., 1963, p. I-1). In these postmodern times this might seem to be an old-fashioned view, but it is one I consider correct.

Experiment plays many roles in science. One of its important roles is to test theories and provide the basis for scientific knowledge. It can also call for a new theory, either by showing that an accepted theory is incorrect or by exhibiting a new phenomenon that is in need of explanation. Experiment can provide hints about the structure or mathematical form of a theory, and it can provide evidence for the existence of the entities involved in our theories. Finally, it may also have a life of its own, independent of theory: Scientists may investigate a phenomenon just because it looks interesting. Such experiments may provide evidence for future theories to explain.

In all of this activity, however, we must remember that science is fallible. Theoretical calculations, experimental results, or the comparison between experiment and theory may all be wrong. Science is more complex than "The scientist proposes, Nature disposes." It may not always be clear what the scientist is proposing. Theories must often be articulated and clarified. It also may not be clear just how nature is disposing. Experiments may not always give clear-cut results, and they may even disagree for a time. Sometimes they can be incorrect.

If experiment is to play these important roles in science, then we must have good reasons to believe experimental results. I present here an epistemology of experiment, a set of strategies that provides reasonable belief in experimental results. Scientific knowledge can then be reasonably based on these experimental results.

Not everyone agrees. Harry Collins, for that example, remarks that "the natural world has a small or non-existent role in the construction of scientific knowledge" (Collins, 1981, p. 3).[1] And Barry Barnes has stated that "Reality will tolerate alternative descriptions without protest. We may say what we will of it, and it will not disagree. Sociologists of knowledge rightly reject epistemologies that *empower* reality" (Barnes, 1991, p. 331).[2] This view led Andy Pickering to remark that "there is no obligation upon anyone framing a view of the world to take account of what twentieth-century science has to say."[3] In this book I argue for the view that nature, as revealed by experiment, plays an important and legitimate role in science. I will begin by offering my own version of an epistemology of experiment, a set of strategies used by scientists to argue for the correctness of an experimental result. I have argued elsewhere that such strategies are justified. I also discuss the views of other scholars, some that support my own view and others that do not.

Experimental Results

The Case for Learning from Experiment

AN EPISTEMOLOGY OF EXPERIMENT

It has been two decades since Ian Hacking asked "Do we see through a microscope?" (Hacking, 1981). Hacking's question really asked How do we come to believe in an experimental result obtained with a complex experimental apparatus? How do we distinguish between a valid result[4] and an artifact created by that apparatus? If experiment is to play all of the important roles in science mentioned above and to provide the evidential basis for scientific knowledge, then we must have good reasons to believe in those results. Hacking (1983) provided an extended answer in the second half of *Representing and Intervening*. He pointed out that even though an experimental apparatus is laden with (at the very least) the theory of the apparatus, observations remain robust despite changes in the theory of the apparatus or the theory of the phenomenon. His illustration was the sustained belief in microscope images despite the major change in the theory of the microscope when Abbe pointed out the importance of diffraction in its operation. One reason Hacking gave for this continued belief is that in making such observations, the experimenters intervened—they manipulated the object under observation. Thus, in looking at a cell through a microscope, one might inject fluid into the cell or stain the specimen. One expects the cell to change shape or color when this is done. Observing the predicted effect strengthens our belief in the proper operation of the experimental apparatus and in the validity of the observation itself.

Hacking also discussed the strengthening of belief in an observation by independent confirmation. The fact that the same pattern of dots—dense bodies in cells—is seen with "different" microscopes (e.g., ordinary, polarizing, phase-contrast, fluorescence, interference, electron, acoustic) argues for the validity of the observation. One might question whether "different" is a theory-laden term. After all, it is our theories of light and the microscope that allow us to consider these microscopes as different from each other. Nevertheless, the argument holds: Hacking correctly argues that it would be a preposterous coincidence if the same pattern of dots were produced in two totally different kinds of physical systems. Different apparatuses have different backgrounds and systematic errors, making the coincidence, if it is an artifact, most unlikely. If it is a correct result, and the instruments are working properly, the agreement of results is understandable.[5]

Hacking's answer is correct as far as it goes. It is, however, incomplete. What happens when one can perform the experiment with only one type of apparatus, such as an electron microscope or a radio telescope, or when intervention is either impossible or extremely difficult? Other strategies are needed to validate the observation.[6] These may include:

1. Experimental checks and calibration, in which the experimental apparatus reproduces known phenomena. For example, if we wish to argue that the spectrum of a substance obtained with a new type of

spectrometer is correct, we might check that this new spectrometer could reproduce the known Balmer series for hydrogen. If we correctly observe the Balmer series, then we strengthen our belief that the spectrometer is working properly. This also strengthens our belief in the results obtained with that spectrometer. If the check fails, then we have good reason to question the results obtained with that apparatus.[7]

2. Reproducing artifacts that are known in advance to be present. An example of this comes from experiments to measure the infrared spectra of organic molecules (Randall et al., 1949). It was not always possible to prepare a pure sample of such material. Sometimes the experimenters had to place the substance in an oil paste or in solution. In such cases, one expects to observe the spectrum of the oil or the solvent superimposed on that of the substance; one can then compare the composite spectrum with the known spectrum of the oil or the solvent. Observation of this artifact gives confidence in other measurements made with the spectrometer.

3. Elimination of plausible sources of error and alternative explanations of the result (the Sherlock Holmes strategy).[8] Thus, when scientists claimed to have observed electric discharges in the rings of Saturn, they argued for their result by showing that it could not have been caused by defects in the telemetry, interaction with the environment of Saturn, lightning, or dust. The only remaining explanation of their result was that it was due to electric discharges in the rings—there was no other plausible explanation of the observation. (In addition, the same result was observed by both spacecrafts Voyager 1 and Voyager 2. This provided independent confirmation. Often, several epistemological strategies are used in the same experiment.)

4. Using the results themselves to argue for their validity. Consider the problem of Galileo's telescopic observations of the moons of Jupiter. Although one might very well believe that his primitive, early telescope might have produced spurious spots of light, it is extremely implausible that the telescope would create images that would appear to be eclipses and other phenomena consistent with the motions of a small planetary system. It is even more implausible that the created spots would satisfy Kepler's Third Law ($R^3/T^2 = $ constant).[9] A similar argument was used by Robert Millikan to support his observation of

the quantization of electric charge and his measurement of the charge of the electron. Millikan (1911) remarked, "The total number of changes which we have observed would be between one and two thousand, and *in not one single instance has there been any change which did not represent the advent upon the drop of one definite invariable quantity of electricity or a very small multiple of that quantity*" (p. 360). In both of these cases one is arguing that there was no plausible malfunction of the apparatus (or no confounding background) that would explain the observations.

5. Using an independently well-corroborated theory of the phenomena to explain the results. This was illustrated in the discovery of the $W^{\pm}$, the charged intermediate vector boson required by the Weinberg-Salam unified theory of electroweak interactions. Although these experiments used very complex apparatuses and used other epistemological strategies (for details, see Franklin [1986], pp. 170–72). I believe that the agreement of the observations with the theoretical predictions of the particle properties helped to validate the experimental results. In this case, the particle candidates were observed in events that contained an electron with high transverse momentum and in which there were no particle jets, just as predicted by the theory. In addition, the measured particle mass of 81 ± 5 GeV/c^2 and 80^{+10}_{-6}, GeV/c^2, found in the two experiments (note the independent confirmation), was in good agreement with the theoretical prediction of 82 ± 2.4 GeV/c^2. It was very improbable that any background effect, which might mimic the presence of the particle, would be in good agreement with theory.

6. Using an apparatus based on a well-corroborated theory. In this case, the support for the theory inspires confidence in the apparatus based on that theory. This is the case with the electron microscope and the radio telescope, whose operations are based on well-supported theories, although other strategies are also used to validate observations made with these instruments.

7. Using statistical arguments. An interesting example of this arose in the 1960s, when the search for new particles and resonances occupied a substantial fraction of the time and effort of those physicists working in experimental high-energy physics. The usual technique was to plot the number of events observed as a function of the invariant

mass of the final-state particles and to look for bumps above a smooth background. The usual informal criterion for the presence of a new particle was that it resulted in a three-standard-deviation effect above the background, a result that had a probability of 0.27% of occurring in a single bin. This criterion was later changed to four standard deviations, which had a probability of 0.0064% when it was pointed out that the number of graphs plotted each year by high-energy physicists made it rather probable, on statistical grounds, that a three-standard-deviation effect would be observed.[10]

These strategies, along with Hacking's intervention and independent confirmation, constitute an epistemology of experiment: They provide us with good reasons for belief in experimental results. They do not, however, guarantee that the results are correct. There are many experiments in which these strategies are applied, but whose results are later shown to be incorrect (examples are presented throughout this book). Experiment is fallible. Neither are these strategies exclusive or exhaustive. No single one of them, or fixed combination of them, guarantees the validity of an experimental result. As the episodes discussed in this book show, physicists use as many of the strategies as they can conveniently apply in any given experiment.

GALISON'S ELABORATION

In *How Experiments End*, Peter Galison (1987) extended the discussion of experiment to more complex situations. In his histories of the measurements of the gyromagnetic ratio of the electron, the discovery of the muon, and the discovery of weak neutral currents, he considers a series of experiments measuring a single quantity, a set of different experiments culminating in a discovery, and two high-energy physics experiments performed by large groups with complex experimental apparatus.

Galison's view is that experiments end when the experimenters believe that they have a result that will stand up in court—a result that I believe includes the use of the epistemological strategies discussed earlier. Thus, David Cline, one of the weak neutral-current experimenters remarked, "At present I don't see how to make these effects [the weak neutral-current event candidates] go away" (Galison, 1987, p. 235).

Galison emphasizes that, within a large experimental group, different members of the group may find different pieces of evidence most convincing. Thus, in the Gargamelle weak neutral-current experiment, several group members found the single photograph of a neutrino-electron scattering event particularly important, whereas for others, the difference in spatial distribution between the observed neutral-current candidates and the neutron background was decisive. Galison attributes this, in large part, to differences in experimental traditions, in which scientists develop skill in using certain types of instruments or apparatuses. In particle physics, for example, there is the tradition of visual detectors, such as the cloud chamber or the bubble chamber, in contrast to the electronic tradition of Geiger and scintillation counters and spark chambers. According to Galison, scientists within the visual tradition tend to prefer "golden events" that clearly demonstrate the phenomenon in question, whereas those in the electronic tradition tend to find statistical arguments more persuasive and important than individual events. (For further discussion of this issue, see Galison [1997] and the next section.)

Galison points out that major changes in theory and in experimental practice and instruments do not necessarily occur at the same time. This persistence of experimental results provides continuity across conceptual changes. Thus, the experiments on the gyromagnetic ratio spanned classical electromagnetism, Bohr's old quantum theory, and the new quantum mechanics of Heisenberg and Schrodinger. Robert Ackermann (1985) has offered a similar view in his discussion of scientific instruments:

The advantages of a scientific instrument are that it cannot change theories. Instruments embody theories, to be sure, or we wouldn't have any grasp of the significance of their operation. . . . Instruments create an invariant relationship between their operations and the world, at least when we abstract from the expertise involved in their correct use. When our theories change, we may conceive of the significance of the instrument and the world with which it is interacting differently, and the datum of an instrument may change in significance, but the datum can nonetheless stay the same, and will typically be expected to do so. An instrument reads 2 when exposed to some phenomenon. After a change in theory,[11] it will continue to show the same reading, even though we may take the reading to be no longer important, or to tell us something other than what we thought originally. (p. 33)

Galison also discusses other aspects of the interaction between experiment and theory. Theory may influence what is considered to be a real effect, demanding explanation, and what is considered background. In his discussion of the discovery of the muon, he argues that the calculation of Oppenheimer and Carlson, which showed that showers were to be expected in the passage of electrons through matter, left the penetrating particles, later shown to be muons, as the unexplained phenomenon. Prior to their work, physicists thought the showering particles were the problem, whereas the penetrating particles seemed to be understood.

The role of theory as an "enabling theory" (i.e., one that allows calculation or estimation of the size of the expected effect and the size of expected backgrounds) is also discussed by Galison (see also Franklin [1995b]). Such a theory can help to determine whether an experiment is feasible. Galison emphasizes that elimination of background that might simulate or mask an effect is central to the experimental enterprise, and not just a peripheral activity. In the case of the weak neutral-current experiments, the existence of the currents depended crucially on showing that the event candidates could not all be due to neutron background.[12]

There is also a danger that the design of an experiment may preclude observation of a phenomenon. Galison points out that the original design of one of the neutral current experiments, which included a muon trigger, would not have allowed the observation of neutral currents. In its original form, the experiment was designed to observe charged currents, which produce a high-energy muon. Neutral currents do not. Therefore, having a muon trigger precluded their observation. Only after the theoretical importance of the search for neutral currents was emphasized to the experimenters was the trigger changed. Changing the design did not, of course, guarantee that neutral currents would be observed.

Galison shows that the theoretical presuppositions of the experimenters may enter into the decision to end an experiment and report the result. Einstein and de Haas ended their search for systematic errors when their value for the gyromagnetic ratio of the electron, $g = 1$, agreed with their theoretical model of orbiting electrons. This effect of presuppositions might cause one to be skeptical of both experimental results and their role in theory evaluation. Galison's history shows, however, that, in this case, the importance of the measurement led to many repetitions of the measurement. This resulted in an agreed-upon result that diverged from theoretical expectations: Scientists do not always find what they are looking for.

STALEY VERSUS GALISON

Recently, Galison has modified his views. In *Image and Logic,* an extended study of instrumentation in 20th-century high-energy physics, Galison (1997) has extended his argument that there are two distinct experimental traditions within that field—the visual (or image) tradition and the electronic (or logic) tradition. The image tradition uses detectors such as cloud chambers or bubble chambers, which provide detailed and extensive information about each individual event. The electronic detectors used by the logic tradition, such as Geiger counters, scintillation counters, and spark chambers, provide less detailed information about individual events, but detect more events. Galison's view is that experimenters working in these two traditions form distinct epistemic and linguistic groups that rely on different forms of argument.[13] The visual tradition emphasizes the single "golden" event. "On the image side resides a deep-seated commitment to the production of the 'golden event': the single picture of such clarity and distinctness that it commands acceptance" (Galison, 1997, p. 22). "The golden event was the exemplar of the image tradition: an individual instance so complete, so well defined, so 'manifestly' free of distortion and background that no further data had to be invoked" (p. 23). Because the individual events provided in the logic detectors contained less detailed information than the pictures of the visual tradition, statistical arguments based on large numbers of events were required.[14]

Kent Staley (1999) disagrees. He argues that the two traditions are not as distinct as Galison believes:

I show that discoveries in both traditions have employed the same statistical [I would add "and/or probabilistic"] form of argument, even when basing discovery claims on single, golden events. Where Galison sees an epistemic divide between two communities that can only be bridged by a creole- or pidgin-like 'interlanguage,' there is in fact a shared commitment to a statistical form of experimental argument. (p. 196).

Staley believes that although there is certainly epistemic continuity within a given tradition, there is also a continuity between the traditions. This does not, I believe, mean that the shared commitment comprises all of the arguments offered in any particular instance, but rather that the same methods are often used by both communities. Galison does not deny that statistical methods are used in the image tradition, but he thinks that they are relatively unimportant. "While statistics could certainly be used within the image tradition, it was by no means necessary for most applications" (Galison 1997, p. 451). In contrast, Galison believes that experiments in the logic tradition "were inherently and inalienably statistical. Estimation of probable errors and the statistical excess over background is not a side issue in these detectors—it is central to the possibility of any demonstration at all" (p. 451). As we shall see, Galison himself presents an example from the visual tradition that exemplifies the use of statistical strategies.

It is interesting to examine the disagreement between Staley and Galison because it illuminates and illustrates issues in the epistemology of experiment. This examination will also show the complexity of demonstrating the validity of an experimental result and the care shown in that demonstration.[15] I will begin with a discussion of what they both regard as a golden event:[16] Anderson's photograph that provided evidence for the existence of the positron (Figure I.1).

The image in question is a cloud chamber photograph that shows two tracks, one on either side of a 6 mm lead plate inserted into the chamber. The two tracks match up very closely, suggesting a single particle passing through the lead. Differences in the curvatures of the tracks above and below indicate a higher energy below the lead than above, which entails, on the assumption that it is indeed a single particle and that particles do not *gain* energy when passing through lead, that the particle was traveling from the lower to the upper region of the space in the photograph. Knowing the direction and curvature of the path, as well as the magnetic field, Anderson concludes that the particle has a positive charge. But based on the length of the track and the energy indicated by the curvature, it cannot have been a proton, which would have had a much shorter range. The particle, then, must have much lighter mass, on the same order of magnitude as that of a free negative electron. (Staley, 1999, p. 215)[17]

Staley argues that Anderson was, in fact, making a statistical argument premised on the claim that the probability of a background event that might have mimicked the presence of a positron was small even when compared to the single event under consideration. Anderson explicitly makes such an argument. In considering alternative explanations of the photograph demonstrating the existence of the positron, he stated:

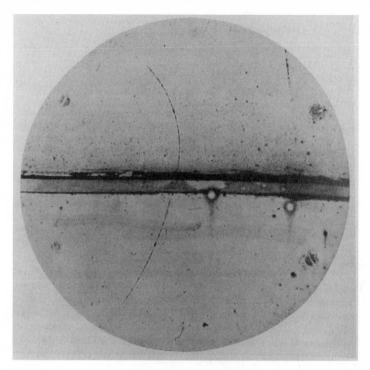

Figure I.1. Anderson's "golden event." The original caption for this figure reads, "A 63 million volt positron ($H\rho = 2.1 \times 10^5$ gauss-cm) passing through a 6 mm lead plate emerging as a 23 million volt positron. [The positron is traveling toward the top of the figure.] The length of this latter path is at least ten times greater than the possible length of a proton path of this curvature." From Anderson (1933, p. 492).

The only escape from this conclusion would be to assume that at exactly the same instant (and the sharpness of the tracks determines that instant to within about a fiftieth of a second) two independent electrons happened to produce two tracks so placed as to give the impression of a single particle shooting through the lead plate. This assumption was dismissed on a *probability basis*, since a sharp track of this order of curvature under the experimental conditions prevailing occurred in the chamber only once in some 500 exposures, and since there was practically no chance at all that two such tracks should line up in this way. (Anderson 1933, p. 491, emphasis added)

As Staley notes, if the probability of a single track is one in 500 exposures, the probability of two such tracks in the same photograph is one in 250,000 exposures, and the probability that they would line up so as to appear to be a single track reduces this probability even further. This was a negligible background indeed, considering the fact that Anderson had only 1,300 exposures.

Galison (1999), in response, notes that Anderson considered four alternative explanations of the photograph:

1. Light positive particle penetrated the lead (ionization ruled out a proton).

2. Simultaneous ejection of positron and electron.

3. Electron *gained* energy in passing downwards through the lead.

4. Two independent electron tracks were perfectly aligned to imitate a positron losing energy. (p. 272)

Galison notes that the first two posit the existence of the positron and thus are not alternative explanations and the third is ruled out by energy conservation.[18] He then asks "Why promote 4) to being *the* unifying epistemological basis of the discovery?" (p. 272). In my view, Galison's "*the*" is an exaggeration. Staley does mention the other alternatives and is here showing that in this particular golden event, the experimenter could, and did, argue on statistical or probabilistic grounds that the background was negligible, and thus that the observation was a real effect. Staley's analysis shows that statistical arguments were one of the arguments used by those in the visual tradition. He is not claiming that it is always the sole argument, or that it is always used. Galison correctly points out that the golden event can be and has been decisive in many instances. He cites Powell et al. (1959): "It is a remarkable feature of those methods in nuclear physics based on recording individual tracks, that the observation of a single event has frequently been of decisive importance in leading to the discovery of phenomena of fundamental importance." Note, however, that Powell and company say "frequently," not "always." What we have seen here is an example of the Sherlock Holmes strategy, in which the elimination of alternative explanations of an experimental result involved the use of statistical arguments.

Staley presents other arguments supporting his view that statistical arguments are not only used within the image tradition, but are often of crucial importance. He presents a discussion of the episode of the discovery of the η meson. In this episode, bubble chamber photographs of the interaction of π^+ mesons with deuterium were examined. Events fitting the hypothesis $\pi^+ + d \rightarrow p + p + \pi^+ + \pi^- + \pi^0$ were analyzed and

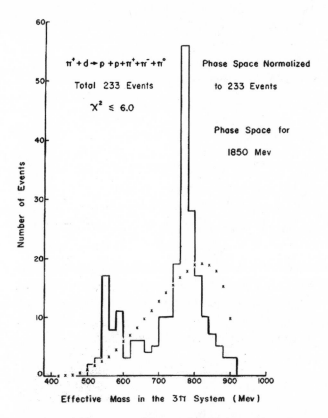

$\pi^+ + d \rightarrow p + p + \pi^+ + \pi^- + \pi^0$

Total 233 Events

$\chi^2 \leq 6.0$

Phase Space Normalized

to 233 Events

Phase Space for

1850 Mev

Figure I.2. A histogram of the number of events plotted against the invariant mass of the three-pion system. The large peak at 770 MeV is due to the known ω^0 meson. The smaller peak at 550 MeV is the suggested new η particle. The phase space distribution is indicated by the x's. From Pevsner et al. (1961).

the number of events as a function of the invariant mass of the $\pi^+\pi^-\pi^0$ system were plotted (Figure I.2).[19] Staley shows that statistical arguments were not only used in the identification of the events, but were also used to establish that the peak at 550 MeV was not due to a statistical fluctuation in the background. (The peak at 770 MeV was caused by the known ω^0 meson.) The experimenters calculated the background expected if the events were distributed according to the phase space available and gave a statistical argument for the presence of the η meson:

We have calculated the Lorentz-invariant phase space for 3-pion mass from the background reaction . . . using the experimental average of the total energy in the p-3π center-of-mass system.

Clearly, because of the presence of the ω^0 particle at 770 MeV, such a normalization of phase space yields a gross overestimate of events expected near 550 MeV. Between 540 and 600 MeV there are 36 events in the experimental distribution, whereas the overestimated phase space would account for 12. (Pevsner et al., 1961, p. 422)

This is actually an application of the statistical strategy included in the earlier discussion of the epistemology of experiment. In this case, although no quantitative estimate of the statistical significance of the proposed signal was made, it is clearly a statistical argument. Staley (1999) concludes and Galison agrees that "Whatever marks the distinction between statistical and non-statistical arguments, it cannot simply be the form of the data" (p. 207). Staley goes on to note "that the prevalence of statistical arguments based on bubble chamber data should already provide grounds for wondering whether the lines are being drawn correctly" (pp. 207–8). This activity (known in the high-energy community as "bump hunting") was, I believe, a considerable and significant fraction of the work done by the bubble-chamber community (visual tradition) during the 1960s and 1970s.[20]

Staley also discusses an event from a logic-tradition experiment that was almost a golden event:[21] Blas Cabrera's "magnetic monopole" (Cabrera, 1982). Magnetic monopoles, whose existence has never been successfully demonstrated, were first introduced by Dirac as a possible explanation for charge quantization. Modern experiments have searched for monopoles by looking for changes in magnetic flux in a superconducting loop. For a single loop, the change in flux was predicted to be $2\varphi_0$, where $\varphi_0 = hc/2e$, where h is Planck's constant, c is the speed of light, and e is the charge of the electron. This was the method used by Cabrera. He used a four-turn superconducting loop, and in 151 days of running, observed one event with a flux change of $8\varphi_0$, exactly the flux change that Dirac had predicted.

Cabrera, however, made no discovery claim based on this single event. He considered various sources of background that might mimic the presence of a magnetic monopole. These included line voltage fluctuations, radiofrequency interference from the rotor brushes of a heat gun, external magnetic field changes, ferromagnetic contamination, the superconducting loop going critical, seismic disturbances, and energetic cosmic rays. Each of these was eliminated (an example of the Sherlock

Holmes strategy), and only one other plausible source of background remained: the possibility of mechanically produced offsets. This was investigated by "sharp raps with a screwdriver handle against the detector assembly." Two out of the 25 blows produced offsets in excess of $6\varphi_0$ (these offsets were followed by drifts in the detector output that were not present in the monopole candidate event). Such a mechanical effect was "not seen as a possible cause for the event," but Cabrera admitted that it could not be ruled out. It was precisely because he could not eliminate this last source of background that Cabrera made no claim that he had observed a monopole. In a comment to Staley he stated, "It was a striking event, because it was exactly the right step size. I thought that there was a good chance it was caused by magnetic charge, but I was not convinced because of the other possible although improbable mechanism" (Staley 1999, p. 221).

Since Cabrera's initial experiment, both he and others, using even more sensitive detectors for a longer time than the initial experiment, have found no large real *or* spurious signals. This has cast doubt on the original monopole candidate and suggests that real and spurious signals are quite rare or that improvements in the experimental apparatus have eliminated the source of spurious signals.[22]

Galison himself has presented a case in which a bubble-chamber experiment used statistics to establish a golden event (Galison, 1987, Chapter 4): the Gargamelle heavy-liquid bubble-chamber experiment that demonstrated the existence of weak neutral currents. What makes this episode so interesting is that two different subgroups of the experimental group used very different methods to search for the phenomenon in question. The Gargamelle bubble chamber was exposed to a beam of muon neutrinos, and the first method used to demonstrate the existence of weak neutral currents was to attempt to show that hadron (strongly interacting particles) showers not containing a muon were produced. The problem was that such showers could also be produced by neutrons. Thus, one had to demonstrate that there was an excess of showers over the number of neutron-induced events. The subgroup used a sample of associated events, neutron-induced hadron showers produced by neutrons generated in charged-current events that were also visible in the chamber. (Charged-current events, already well established, contained a hadron shower and a muon. The neutral-current events had no muons). Using statistical techniques and computer simulations, the experimenters found that neutron-induced background could account for only 20% of the neutral-current candidates.[23] In addition, the flat distribution of the candidate events did not match that expected for neutron-induced events, which would peak near the front of the chamber: Neutral currents had been observed.

The second subgroup searched for neutral currents by looking for examples of neutrino-electron scattering in the same bubble-chamber exposure. Such events could only arise if neutral currents existed. They found one such golden event: "this event was a 'Bilderbuch example' of what we had been expecting [for] months to show up: a candidate for neutrino electron scattering. *But the crucial point was to assess background*" (Helmut Faissner, quoted in Galison [1987], p. 181, emphasis added). The most obvious background was electrons from ordinary inverse β decay, $\nu_e + n \rightarrow e + p$, in which the proton wasn't observed. This background was estimated using charged-current events of the form $\nu_\mu + n \rightarrow \mu + p$. The experimenters found that the ratio (Hard muons without observed protons)/(Hard muons with observed protons) = 0.03 ± 0.02. It was also known that electron and muon charged-current interactions were identical at these energies, so the same ratio applied to electron events. Using the number of electrons observed with protons and the relative number of electron- and muon neutrinos in the beam, the experimenters calculated a background of 0.09 ± 0.07 events. This background level made the observation of one event extremely unlikely.

Another possibility was asymmetric electron-positron pair production, in which only the electron was observed. Using the number of observed pairs (only one), the background due to this effect was 0.015 events. Another possible background was Compton scattering, in which the observed electron was produced by a γ ray. "But the Aachen electron was so energetic that this possibility never even arose at the collaboration meeting" (Galison 1987, p. 184). A calculation confirmed that the background was indeed small: Using the calculated ratio of Compton scattering to pair production (0.5%) and the one observed pair event gave a background of 0.005 events. The observed neutrino-electron scattering event was indeed golden, but only after it was shown that it could not be due to background.

Despite their differences over whether the image and logic traditions consist of different linguistic and epistemic communities, [24] both Staley and Galison have presented evidence that we learn from experiment and that that knowledge is supported by good reasons.

The Case against Learning from Experiment

COLLINS AND THE EXPERIMENTERS' REGRESS

Collins, Pickering, and others have raised objections to the view that experimental results are accepted on the basis of epistemological arguments. As Donald MacKenzie (1989), for example, remarks:

Recent sociology of science, following sympathetic tendencies in the history and philosophy of science, has shown that no experiment, or set of experiments however large, can on its own compel resolution of a point of controversy, or, more generally acceptance of a particular fact. A sufficiently determined critic can always find reason to dispute any alleged "result." If the point at issue is, say, the validity of a particular theoretical claim, those who wish to contest an experimental proof or disproof of the claim can always point to the multitude of auxiliary hypotheses (for example about the operation of instruments) involved in drawing deductions from a given theoretical statement to a particular experimental situation or situations. One of these auxiliary hypotheses may be faulty, critics can argue, rather than the theoretical claim apparently being tested. Further the validity of the experimental procedure can also be attacked in many ways. (p. 412)

MacKenzie is raising doubts not only about the validity of experimental results, but also on their use in testing theories or hypotheses. I will begin with the former. There are two points at issue. The first involves the meaning one assigns to "compel." If one reads it, as Mackenzie seems to, as "entail," then I agree that no finite set of confirming instances can entail a universal statement. No matter how many white swans one observes it does not entail that "all swans are white." Neither can any argument, no matter how persuasive or valid, establish with absolute certainty the correctness of an experimental result. A more reasonable meaning for "compel" is having good reasons for belief. As the episodes discussed in this book demonstrate, this is the meaning used in science, and those reasons for belief in an experimental result are provided by the epistemology of experiment.

The second point is a logical one, known to philosophers of science as the Duhem-Quine problem (see Harding, 1976). In the usual *modus tollens,* if a hypothesis *h* entails an experimental result *e* then ¬*e* (not *e*) entails ¬*h*. As Duhem and Quine both pointed out, it is not just *h* that entails *e* but rather *h* and *b*, where *b* includes background knowledge and auxiliary hypotheses. Thus, ¬*e* entails ¬*h* or ¬*b* and we do not know where to place the blame.[25] I am assuming here a weak form of the Duhem-Quine problem, in which one assumes the experimental result ¬*e* is correct. One can, of course, as MacKenzie does, challenge that experimental result. As Quine pointed out, any statement can be maintained come what may, provided one is willing to make changes elsewhere in one's background knowledge. The question is when the price that one has to pay becomes too high to justify maintaining the statement. As we see in this book, individual scientists can maintain a belief in their own results, despite arguments that the rest of the physics community find convincing. Thus, for example, Weber never gave up his belief that he had found evidence for gravity waves. In almost all cases, however, even the proponents of a particular theory or the physicists who reported an experimental result are persuaded by reasonable epistemological, methodological, and evidential arguments that their views are incorrect. Thus, Simpson no longer believes in the existence of the 17-keV neutrino, and Fischbach, Aronson, and Talmadge no longer believe in the existence of a "Fifth Force" in gravity. Several other examples are also presented in later chapters.

MacKenzie's skepticism illustrates one of the underlying principles of what has been called the sociology of scientific knowledge. Advocates of that view argue that because experimental evidence or methodological rules cannot resolve points of controversy, other reasons must be invoked to explain the resolution and those reasons are social.

Harry Collins, for example, is well known for his skepticism concerning both experimental results and evidence. Collins (1985, pp. 79–111) develops an argument that he calls the "experimenters' regress": What scientists take to be a correct result is one obtained with a good, that is, properly functioning, experimental apparatus. But a good experimental apparatus is simply one that gives correct results. Collins claims that there are no formal criteria that one can apply to decide whether an experimental apparatus is working properly. In particular, he argues that calibrating an experimental apparatus by using a surrogate signal cannot provide an independent reason for considering the apparatus to be reliable.

In Collins' view, the regress is eventually broken by negotiation within the appropriate scientific community, a process driven by factors such as the career, social, and cognitive interests of the scientists, and the perceived utility for future work, but one that is not decided by what we might call epistemological criteria, or reasoned judgment. Thus Collins concludes that his regress raises serious questions concerning both experimental evidence and its use in the evaluation of scientific hypotheses and theories. Indeed, if no way out of the regress can be found, then he has a point.

Collins' strongest candidate for an example of the experimenters' regress is presented in his history of the early attempts to detect gravitational radiation, or gravity waves. (For more detailed discussion of this episode, see Collins [1985, 1994]; Franklin [1994, 1997a].) This episode will be discussed in detail in Chapter 2 and I will present only a brief summary here. In this case, the physics community was forced to compare Weber's claims that he had observed gravity waves with the reports from six other experiments that failed to detect them. On the one hand, Collins argues that the decision between these conflicting experimental results could not be made on epistemological or methodological grounds—he claims that the six negative experiments could not legitimately be regarded as replications[26] and hence become less impressive. On the other hand, Weber's apparatus, precisely because the experiments used a new type of apparatus to try to detect a hitherto unobserved phenomenon,[27] could not be subjected to standard calibration techniques.

Contrary to Collins, I believe that the scientific community made a reasoned judgment when rejecting Weber's results and accepting those of his critics. Although no formal rules were applied (e.g., if you make four errors, rather than three, your results lack credibility; or if there are five, but not six, conflicting results, your work is still credible), the procedure was reasonable.

PICKERING: COMMUNAL OPPORTUNISM AND PLASTIC RESOURCES

Pickering has argued that the reasons for accepting results are the future utility of such results for both theoretical and experimental practice and the agreement of such results with the existing community commitments. In discussing the discovery of weak neutral currents, Pickering (1984b) states, "Quite simply, particle physicists accepted the existence of the neutral current because they could see how to ply their trade more profitably in a world in which the neutral current was real" (p. 87). "Scientific communities tend to reject data that conflict with group commitments and, obversely, to adjust their experimental techniques to tune in on phenomena consistent with those commitments" (Pickering, 1981, p. 236). The emphasis on future utility and existing commitments is clear. These two criteria do not necessarily agree. For example, there are episodes in the history of science in which better opportunity for future work is provided by the overthrow of existing theory. (See, for example, the histories of the overthrow of parity conservation and of CP symmetry discussed in Franklin [1986, Chapters 1 and 3].)

Pickering has recently offered a different view of experimental results. In his view, the material procedure (including the experimental apparatus itself along with setting it up, running it, and monitoring its operation), the theoretical model of that apparatus, and the theoretical model of the phenomena under investigation are all plastic resources that the investigator brings into relations of mutual support (Pickering, 1987, 1989). "Achieving such relations of mutual support is, I suggest, the defining characteristic of the successful experiment" (1987, p. 199). He uses Morpurgo's search for free quarks, or fractional charges of 1/3 e or 2/3 e (where e is the charge of the electron) as an example of plastic resources. (See also Gooding [1992].) Morpurgo used a modern Millikan-type apparatus and initially found a continuous distribution of charge values.

Morpurgo began from a conceptual design study of what he believed an adequate charge-measuring device should look like. He set out to implement this design in the material world, to build the apparatus. When the apparatus had been built, he attempted to use it to measure charges (on samples of graphite, initially). And he found that it did not work. Instead of finding integral or fractional charges, he found that his samples appeared to carry charges distributed over a continuum.[28] There followed a period of tinkering, of pragmatic, trial and error, material interaction with the apparatus. This came to an end when Morpurgo discovered that if he increased the separation of capacitor plates within his apparatus he obtained integral charge measurements. . . . After some theoretical analysis, Morpurgo concluded that he now had his apparatus working properly, and reported his failure to find any evidence for fractional charges. . . . Morpurgo would have been happy if his tinkering had eventuated in measurements of fractional charges. In fact, it did not. The point I want to emphasize is that this eventuation was not entirely under Morpurgo's (or anyone's) control; it was a product of Morpurgo's immersion, through the medium of his experiment, in the real. (Pickering, 1987, p. 197)

Pickering goes on to note that Morpurgo did not tinker with the two competing theories of the phenomena then on offer, those of integral and fractional charge:

The initial source of doubt about the adequacy of the early stages of the experiment was precisely the fact that their findings—continuously distributed charges—were consonant with neither of the phenomenal models which Morpurgo was prepared to countenance. And what motivated the search for a new instrumental model was Morpurgo's eventual success in producing findings in accordance with one of the phenomenal models he was willing to accept.

The conclusion of Morpurgo's first series of experiments, then, and the production of the observation report which they sustained, was marked by bringing into relations of mutual support of the three elements I have discussed: the material form of the apparatus and the two conceptual models, one instrumental and the other phenomenal. Achieving such relations of mutual support is, I suggest, the defining characteristic of the successful experiment. (p. 199)

Pickering has made several important and valid points concerning experiment. Most importantly, he has emphasized that an experimental apparatus is rarely initially capable of producing valid experimental results. He has also recognized that both the theory of the apparatus and the theory of the phenomena can enter into the production of a valid experimental result, although I doubt that he would regard these as epistemological strategies. What I wish to question, however, is the emphasis he places on these theoretical components. I have already suggested that the theoretical components can be among the strategies used to argue for the validity of experimental results. I do not believe, as Pickering seems to, that they are necessary parts of such an argument. As Hacking (1983) points out, experimenters had confidence in microscope images before and after Abbe's work fundamentally changed the theoretical understanding of the microscope. This confidence was due to intervention, not theory.

Pickering ignores that prior to Morpurgo's experiment, it was known (or there were at least excellent reasons to believe) that electric charge was quantized in units of e, the charge on the electron, and that fractional charges, if they existed, were very rare in comparison with integral charges. From Millikan onward, experiments had strongly supported the existence of a fundamental unit of charge, and of charge quantization (see Chapter 3). The failure of Morpurgo's apparatus to produce measurements of integral charge indicated that it was not operating properly and that his theoretical understanding of it was faulty. It was the failure to produce measurements in agreement with what was already known (i.e., the failure of an important experimental check) that caused doubts about Morpurgo's measurements. This was true regardless of the theoretical models available, or those that Morpurgo was willing to accept. It was only when Morpurgo's apparatus could reproduce known measurements that it could be trusted and used to search for fractional charge.[29] To be sure, Pickering has allowed a role for the real in the production of the experimental result, but it does not seem to be decisive. I have argued that it is.

CRITICAL RESPONSES TO PICKERING

Ackermann has offered a modification of Pickering's view. Ackermann (1991) suggests that the experimental apparatus itself is a less plastic resource than the theoretical model of the apparatus or that of the phenomenon:

To repeat, changes in A [the apparatus] can often be seen (in real time, without waiting for accommodation by B [the theoretical model of the apparatus]) as improvements, whereas "improvements" in B don't begin to count unless A is actually altered and realizes the improvements conjectured. It's conceivable that this small asymmetry can account, ultimately, for large scale directions of scientific progress and for the objectivity and rationality of those directions. (p. 456)

Hacking (1992) has also offered a more complex version of Pickering's later view. He suggests that the results of mature laboratory science achieve stability and are self-vindicating when the elements of laboratory science are brought into mutual consistency and support. These are (1) ideas: questions, background knowledge, systematic theory, topical hypotheses, and modeling of the apparatus; (2) things: target, source of modification, detectors, tools, and data generators; and (3) marks and the manipulation of marks: data, data assessment, data reduction, data analysis, and interpretation. "Stable laboratory science arises when theories and laboratory equipment evolve in such a way that they match each other and are mutually self-vindicating" (Hacking, 1992, p. 56). "We invent devices that produce data and isolate or create phenomena, and a network of different levels of theory is true to these phenomena. Conversely we may in the end count them only as phenomena only when the data can be interpreted by theory" (pp. 57–58). One might ask whether such mutual adjustment between theory and experimental results can always be achieved. What happens when an experimental result is produced by an apparatus on which several of the epistemological strategies discussed earlier have been successfully applied, and the result is in disagreement with our theory of the phenomenon? Accepted theories can be refuted. (See Franklin, 1986, Chapters 1 and 3.)

Hacking himself worries about what happens when a laboratory science that is true to the phenomena generated in the laboratory, thanks to mutual adjustment and self-vindication, is successfully applied to the world outside the laboratory. Does this argue for the truth of the science? In Hacking's view it does not. If laboratory science does produce happy effects in the "untamed world . . . it is not the truth of anything that causes or explains the happy effects" (Hacking, 1992, p. 60).

PICKERING AND THE DANCE OF AGENCY

Recently Pickering (1995) has offered a somewhat revised account of science. "My basic image of science is a performative one, in which the performances—the doings—of human and material agency come to the fore. Scientists are human agents in a field of material agency which they struggle to capture in machines" (p. 21). He then discusses the complex interaction between human and material agency, which I interpret as the interaction between experimenters, their apparatus, and the natural world:

The dance of agency, seen asymmetrically from the human end, thus takes the form of a *dialectic of resistance and accommodations,* where resistance denotes the failure to achieve an intended capture of agency in practice, and accommodation an active human strategy of response to resistance, which can include revisions to goals and intentions as well as to the material form of the machine in question and to the human frame of gestures and social relations that surround it. (p. 22)

Pickering's idea of resistance is illustrated by Morpurgo's observation of continuous, rather than integral or fractional, electrical charge, which did not agree with his expectations. Morpurgo's accommodation consisted of changing his experimental apparatus by using a larger separation between his plates, and also by modifying his theoretical account of the apparatus. That being done, integral charges were observed and the result stabilized by the mutual agreement of the apparatus, the theory of the apparatus, and the theory of the phenomenon. Pickering notes that "the outcomes depend on how the world is" (p. 182). "In this way, then, *how the material world is* leaks into and infects our representations of it in a nontrivial and consequential fashion. My analysis thus displays an intimate and responsive engagement between scientific knowledge and the material world that is integral to scientific practice" (p. 183).

Nevertheless there is something confusing about Pickering's invocation of the natural world. Although Pickering acknowledges the importance of the natural world, his use of the term "infects" seems to indicate that he is not entirely happy with this. Nor does the natural world seem to have much efficacy. It never seems to be decisive in any of Pickering's case studies. Recall that he argued that physicists accepted the existence of weak neutral currents because "they could ply their trade more profitably in a world in which the neutral current was real." In his account, Morpurgo's observation of continuous charge is important only because it disagrees with his theoretical models of the phenomenon. The fact that it disagreed with numerous previous observations of integral charge does not seem to matter. This is further illustrated by Pickering's discussion of the conflict between Morpurgo and Fairbank. As we have seen, Morpurgo reported that he did not observe fractional electrical charges. On the other hand, in the late 1970s and early 1980s, Fairbank and his collaborators published a series of papers in which they claimed to have observed fractional charges (e.g., LaRue et al., 1981). Faced with this discord, Pickering (1995) concludes:

In Chapter 3, I traced out Morpurgo's route to his findings in terms of the particular vectors of cultural extension that he pursued, the particular resistances and accommodations thus precipitated, and the particular interactive stabilizations he achieved. The same could be done, I am sure, in respect of Fairbank. And these tracings are all that needs to be said about their divergence. It just happened that the contingencies of resistance and accommodation worked out differently in the two instances. Differences like these are, I think, continually bubbling up in practice, without any special causes behind them. (pp. 211–12)

The natural world seems to have disappeared from Pickering's account. There is a real question here as to whether fractional charges exist in nature. The conclusions reached by Fairbank and by Morpurgo about their existence cannot both be correct.[30] It seems insufficient merely to state, as Pickering does, that Fairbank and Morpurgo achieved their individual stabilizations and to leave the conflict unresolved. (Pickering does comment that one could follow the subsequent history and see how the conflict was resolved, and he does give some brief statements about it, but its resolution is not important for him.) At the very least, I believe, one should consider the actions of the scientific community. Scientific knowledge is not determined individually, but communally. Pickering seems to acknowledge this. "One might, therefore, want to set up a metric and say that items of scientific knowledge are more or less objective depending on the extent to which they are threaded into the rest of scientific culture, socially stabilized over time, and so on. I can see nothing wrong with thinking this way" (Pickering, 1995, p. 196). The fact that Fairbank believed in the existence of fractional electrical charges, or that Weber strongly believed that he had observed gravity waves, does not make them right: These are questions about the natural world that can be resolved. Either fractional charges and gravity waves exist or they do not, or to be more cautious, we might say that we have good reasons to support our claims about their existence, or we do not.

Another issue neglected by Pickering is the question of whether a particular mutual adjustment of theory (of the apparatus or the phenomenon), the experimental apparatus, and evidence is justified. Pickering seems to believe that any such adjustment that provides stabilization, either for an individual or for the community, is acceptable. I do not think this is correct. As we shall see in Part I, some experimenters both excluded data and engaged in selective analysis procedures in producing experimental results. These practices are, at the very least, questionable, as is the use of the results produced by such practices in science. Consider a simple example. Suppose one wished to show empirically that all odd numbers were prime. One looks at the odd numbers and notes that 1, 3, 5, and 7 are all primes, one excludes 9 as an experimental error, finds 11 and 13 are prime, and then stops looking. Surely no one would, or should, regard this as a legitimate procedure, or base any conclusion on the result! Although this is a rather contrived example, in later chapters I discuss episodes (e.g., the claimed existence of low-mass electron-positron states and the early search for gravity waves) in which similarly questionable procedures occurred.

The difference between our attitudes toward the resolution of discord is one of the important distinctions between my view of science and Pickering's. I do not believe it is sufficient simply to say that the resolution is socially stabilized. I want to know how that resolution was achieved and what were the reasons offered for that resolution. If we are faced with discordant experimental results and both experimenters have offered reasonable arguments for their correctness, then clearly more work is needed. It seems reasonable, in such cases, for the physics community to search for an error in one, or both, of the experiments. Part II of this book is devoted to a detailed examination and discussion of several cases of discordant experimental results and the reasons for the resolution.

Pickering (1995) discusses yet another difference between our views. He sees traditional philosophy of science as regarding objectivity "as stemming from a peculiar kind of mental hygiene or policing of thought. This police function relates specifically to theory choice in science, which . . . is usually discussed in terms of the rational rules or methods responsible for closure in theoretical debate" (p. 197). He goes on to remark that:

The most action in recent methodological thought has centered on attempts like Allan Franklin's to extend the methodological approach to experiments by setting up a set of rules for their proper performance. Franklin thus seeks to extend classical discussions of objectivity to the empirical base of science (a topic hitherto neglected in the philosophical tradition but one that, of course the mangle [Pickering's view] also addresses). For an argument between myself and Franklin on the same lines as that laid out below, see Franklin 1990, Chapter 8; Franklin 1991; and Pickering 1991; and for commentaries related to that debate, Ackermann 1991 and Lynch 1991. (p. 197)

See also Franklin (1993b). Although I agree that my epistemology of experiment is designed to offer good reasons for belief in experimental results, I do not agree with Pickering that they are a set of rules. I regard them as a set of strategies, from which physicists choose, to argue for the correctness of their results. As noted above, I do not think the strategies offered are either exclusive or exhaustive. Judging by Pickering's discussions of Fairbank and Morpurgo and his 1991 essay, he does not think my epistemology of experiment can serve that function.

There is another point of disagreement between Pickering and myself. He claims to be dealing with the practice of science, and yet he excludes certain practices from his discussions. As discussed later (see also Franklin [1986, Chapter 7] for other cases), one scientific practice is the application of the epistemological strategies I have outlined above to argue for the correctness of an experimental result. In fact, one of the essential features of an experimental paper is the presentation of such arguments. I note further that writing such papers, a performative act, is also a scientific practice, and it would seem reasonable to examine both the structure and content of those papers.[31]

Thus, there is significant disagreement on the reasons for the acceptance of experimental results. For some, like Staley, Galison, and myself, it is because of epistemological arguments. For others, like Pickering, the reasons are utility for future practice and agreement with existing theoretical commitments. Although the history of science shows that the overthrow of a well-accepted theory leads to an enormous amount of theoretical and experimental work, proponents of this view seem to accept that it is always agreement with existing theory that has more future utility. Hacking and Pickering also suggest that experimental results are accepted on the basis of the mutual adjustment of elements, including the theory of the phenomenon.

Nevertheless, everyone agrees that a consensus does arise on which experimental results to use.

HACKING'S THE SOCIAL CONSTRUCTION OF WHAT?

Recently Ian Hacking (1999, Chapter 3) has provided an incisive and interesting discussion of the issues that divide the constructivists (e.g., Collins and Pickering) from the rationalists such as myself.[32] He sets out three sticking points between the two views: (1) contingency, (2) nominalism, and (3) external explanations of stability.

Contingency is the idea that science is not predetermined, that it could have developed in any one of several successful ways. This is the view adopted by constructivists. Hacking illustrates this with Pickering's (1984a) account of high-energy physics during the 1970s, when the quark model came to dominate:

The constructionist maintains a *contingency thesis*. In the case of physics, (a) physics (theoretical, experimental, material) could have developed in, for example, a nonquarky way, and, by the detailed standards that would have evolved with this alternative physics, could have been as successful as recent physics has been by its detailed standards.[33] Moreover, (b) there is no sense in which this imagined physics would be equivalent to present physics. The physicist denies that. (Hacking, 1999, pp. 78–79)

To sum up Pickering's doctrine: there could have been a research program as successful ("progressive") as that of high-energy physics in the 1970s, but with different theories, phenomenology, schematic descriptions of apparatus, and apparatus, and with a different, and progressive, series of robust fits between these ingredients. Moreover—and this is something badly in need of clarification—the "different" physics would not have been equivalent to present physics. Not logically incompatible with, just different.

The constructionist about (the idea) of quarks thus claims that the upshot of this process of accommodation and resistance is not fully predetermined. Laboratory work requires that we get a robust fit between apparatus, beliefs about the apparatus, interpretations and analyses of data, and theories. *Before a robust fit has been achieved, it is not determined what that fit will be. Not determined by how the world is, not determined by technology now in existence, not determined by the social practices of scientists, not determined by interests or networks, not determined by genius, not determined by anything.* (pp. 72–73, emphasis added)

As was the case with MacKenzie's use of the term "compel," much depends here on what Hacking means by "determined." If he means entailed, then I agree with him. I doubt that the world, or more properly, what we can learn about it, entails a unique theory. If this is not what he means, as seems more plausible, then he implies that reality places no restrictions on that successful science: I disagree strongly. I would certainly wish to argue that the way the world is restricts the kinds of theories that will fit the phenomena, the kinds of apparatus we can build, and the results we can obtain with such apparatuses. To think otherwise seems silly. Consider another simple example: It seems to me highly unlikely that someone can come up with a successful theory in which objects whose density is greater than that of air fall upwards. This is not, I believe, a caricature of the view Hacking describes. Describing Pickering's view, he states, "Physics did not need to take a route that involved Maxwell's equations, the Second Law of Thermodynamics, or the present values of the velocity of light" (Hacking, 1999, p. 70). Although I have some sympathy for this view as regards Maxwell's Equations or the Second Law of Thermodynamics, I do not agree about the value of the speed of light. That is determined by the way the world is.[34] Any successful theory of light must give that value for its speed.

At the other extreme are the "inevitablists," among whom Hacking classifies most scientists. He cites Sheldon Glashow, a Nobel Prize winner: "Any intelligent alien anywhere would have come upon the same logical system as we have to explain the structure of protons and the nature of supernovae" (Glashow, 1992, p. 28).

Another difference between Pickering and myself on contingency concerns not whether an alternative is possible, but rather whether there are reasons why that alternative should be pursued. Pickering seems to identify *can* with *ought*. This is illustrated in our very different discussions of the episode of atomic parity violation. See Chapter 10 for a detailed discussion of this episode.

Constructivist case studies always seem to result in the support of existing, accepted theory (Pickering, 1984a,b, 1991; Collins, 1985; Collins and Pinch, 1993). One criticism implied in such cases is that alternatives are not considered, that the hypothesis space of acceptable alternatives is either very small or empty. I do not believe this is correct. Thus, when the experiment of Christenson et al. (1964) detected K_2^0 decay into two pions, which seemed to show that CP symmetry (combined particle-antiparticle and space inversion symmetry) was violated, no fewer than 10 alternatives were offered.[35] These included:

1. The cosmological model resulting from the local asymmetry of matter and antimatter;

2. External fields;

3. The decay of the K_2^0 into a K_1^0 with the subsequent decay of the K_1^0 into two pions, which was allowed by the symmetry;

4. The emission of another neutral particle, "the paritino," in the K_2^0 decay, similar to the emission of the neutrino in beta decay;

5. One of the pions emitted in the decay was in fact a "spion," a pion with spin one rather than zero;

6. The decay was due to another neutral particle, the L, produced coherently with the K^0;

7. The existence of a "shadow" universe, which interacted with our universe only through the weak interactions, and that the decay seen was the decay of the "shadow K_2^0";

8. The failure of the exponential decay law;

9. The failure of the principle of superposition in quantum mechanics; and

10. The decay pions were not bosons.

As one can see, the limits placed on alternatives were not very stringent. By the end of 1967, all of the alternatives had been tested and found wanting, leaving CP symmetry unprotected. Here the differing judgments of the scientific community about what was worth proposing and pursuing led to a wide variety of alternatives being tested.[36]

Hacking's second sticking point is nominalism, or name-ism. He notes that in its most extreme form nominalism denies that there is anything in common or peculiar to objects selected by a name, such as "Douglas fir" other than that they are called Douglas fir. Opponents contend that good names, or good accounts of nature, tell us something valid about the world. This is related to the realism-antirealism debate concerning the status of unobservable entities that has plagued philosophers for millennia. For example, Bas van Fraassen (1980), an antirealist, holds that we have no grounds for belief in unobservable entities such as the electron and that accepting theories about the electron means only that we believe that the things the theory says about observables are true.[37] A realist claims that electrons really exist and that as, for example, Wilfred Sellars (1962) remarked, "to have good reason for holding a theory is *ipso facto* to have good reason for holding that the entities postulated by the theory exist" (p. 97). In Hacking's view, a scientific nominalist is more radical than an antirealist and is just as skeptical about fir trees as antirealists are about electrons. A nominalist further believes that the structures we conceive of are properties of our representations of the world and not of the world itself. Hacking refers to opponents of that view as inherent structuralists.

Hacking also remarks that this point is related to the question of "scientific facts." Thus, constructivists such as Latour and Woolgar (1979) originally entitled their book *Laboratory Life: The Social Construction of Scientific Facts*.[38] Andrew Pickering (1984a) entitled his history of the quark model *Constructing Quarks*. Physicists argue that this demeans their work. Steven Weinberg, a realist and a physicist, criticized Pickering's title by noting that no mountaineer would ever name a book *Constructing Everest*. For Weinberg, quarks and Mount Everest have the same ontological status. They are both facts about the world. Hacking argues that constructivists do not, despite appearances, believe that facts do not exist, or that there is no such thing as reality. He cites Latour and Woolgar (1986) "that 'out-there-ness' is a *consequence* of scientific work rather than its cause" (p. 180). I agree with Hacking (1999) when he concludes that:

Latour and Woolgar were surely right. We should not *explain* why some people believe that *p* by saying that *p* is true, or corresponds to a fact, or the facts. For example: someone believes that the universe began with what for brevity we call a big bang. A host of reasons now supports this belief. But after you have listed all the reasons, you should not add, as if it were an additional reason for believing in the big bang, "and it is true that the universe began with a big bang." Or "and it is a fact." This observation has nothing peculiarly to do with social construction. It could equally have been advanced by an old-fashioned philosopher of language. It is a remark about the grammar of the verb "to explain." (pp. 80–81)

I would add, however, that the reasons Hacking cites as supporting that belief are given to us by valid experimental evidence and not by the social and personal interests of scientists. I'm not sure that Latour and Woolgar would agree. My own position is one that one might reasonably call "conjectural realism." I believe that we have good reasons to believe in facts, and in the entities involved in our theories, always remembering, of course, that science is fallible.[39]

Hacking's third sticking point is the external explanations of stability:

The constructionist holds that explanations for the stability of scientific belief involve, at least in part, elements that are external to the content of science. These elements typically include social factors, interests, networks, or however they be described. Opponents hold that whatever be the context of discovery, the explanation of stability is internal to the science itself. (Hacking, 1999, p. 92)

Rationalists think that most science proceeds as it does in the light of good reasons produced by research. Some bodies of knowledge become stable because of the wealth of good theoretical and experimental reasons that can be adduced for them. Constructivists think that the reasons are not decisive for the course of science. Nelson (1994) concludes that this issue will never be decided. Rationalists, at least retrospectively, can always adduce reasons that satisfy *them*. Constructivists, with equal ingenuity, can always find to their own satisfaction an openness where the upshot of research is settled by something other than reason. Something external. That is one way of saying we have found an irresoluble "sticking point." (pp. 91–92)

Hacking seems to agree with Nelson that this is an irresoluble sticking point. (I will have more to say about my disagreement with Nelson in the Conclusion.) Although the adherents of both the strong rationalist and strong constructivist views may be unconvinced (and possibly unconvincible) by each others' accounts, I am unconvinced myself, the majority of those who study science are in the middle and can, I believe, decide for themselves which account is better. Sandra Harding (1996), writing on the so-called "Science Wars" states, "It is significant that the Right's objections virtually never get into the nitty-gritty of historical or ethnographic detail to contest the accuracy of social studies of science accounts. Such objections remain at the level of rhetorical flourishes and ridicule" (p. 15). In a sense Harding is correct, because the number of episodes studied from both a rationalist and constructivist point of view is rather small, but the blame must be shared by both sides.[40] In the discussions in Part II, I present the technical details of several episodes from the history of contemporary physics (some would say too many such details). Some time ago I challenged constructivists to offer alternative accounts of these episodes. That challenge is still unanswered.

Hacking also suggests that one should score oneself on the three sticking points on a scale from 1 to 5, where a score of 5 is a strong constructivist position and 1 is a strong rationalist position. He ranks himself 2 on contingency, 4 on nominalism, and 3 on external explanations of stability.[41] My own scores would be 2, 2, and 1.

One issue that Hacking does not discuss, however, is the relative weight that one assigns to the three sticking points. I rank them, in decreasing order of importance: external explanations of stability, contingency, and nominalism. I believe that one can be a rationalist and not be an inherent structuralist, or a realist, and also consider the issue relatively unimportant. Although I have written on the question of scientific realism (Franklin, 1988, 1996, 1997b; 2000a), I do not regard it as essential to my discussion of science as a reasonable enterprise. I believe that we have good reasons to believe in the entities involved in our theories, but I could give up that view. Similarly a strong rationalist, such as myself, might very well agree that things in science could have been different, that other theories might have been proposed and adopted. They would argue that, whatever view of the world is accepted, it is accepted on the basis of valid experimental evidence. That brings us to sticking point 3: external explanations of stability.

My own view is that this sticking point (external explanations of stability) is one on which I will not compromise. I believe that such explanations are incompatible with the history of science (I present several cases in this book). I also believe that if external factors such as social interests and career interests are crucial in the acceptance of scientific beliefs, then science has, in fact, no claim to knowledge. It is not completely facetious to suggest that on an external view, what is accepted by the scientific community could be determined by a majority vote. Stanley Fish, a well-known cultural critic of science, has likened the laws of nature to the laws of baseball. Professor Fish is clearly mistaken. In major-league baseball there is a designated hitter in the American League, but not in the National League. This was decided on by a vote of the team owners. Not so for the law of gravity: Objects would not fall differently in the United States and in France if the American Physical Society voted to repeal the law of gravity, whereas the French Academy voted to retain it. In my own rough-and-ready view, knowledge is justified belief, and that justification can only be provided by valid experimental evidence and on reasoned and critical discussion.[42]

In the case studies discussed in Part II, I argue that the resolution of discordant results, for example, was settled by methodological and epistemological arguments based on valid experimental evidence. I have found no evidence that social factors played a role here, or in any other episodes I have studied.

Two Problems

Recently, detailed studies of experiment, by myself and by others, have raised two important and serious questions concerning the view I have outlined above. These are the question of selectivity in the production of experimental results and the issue of discordant experimental results. The former involves the application of selection criteria, or "cuts," to either the experimental data or the analysis procedures used to transform that data into an experimental result. One might legitimately question whether the result is valid or is an artifact produced by the cuts. One might also worry about the possibility of experimenter bias in the application of the cuts. This is a particular problem when the effect of the cuts on the final result is known.

In Part I, I outline some solutions to the problem of selectivity and present histories of five episodes that illustrate both the problem and the solutions. I first discuss several strategies used to argue that an experimental result is not an artifact produced by the cuts, and conversely, how one might show that a result is such an artifact. The first episode, the measurement of the K^+_{e2} branching ratio, shows ordinary cuts made on data to produce an experimental result, along with the checks that were performed in order to demonstrate that the result was not an artifact of the cuts. In the second episode, Millikan's measurement of e, the charge of the electron, I show that Millikan engaged in selectivity in both data and in analysis procedures. I also demonstrate that the effects of Millikan's selectivity were quite small. In this episode the correctness of Millikan's result was checked by the subsequent, and numerous, independent measurements of e.

In each of these episodes the result was shown to be correct. This was not the case in the other three episodes—the early search for gravity waves, the claimed existence of a heavy, 17-keV neutrino, and the claimed existence of low-mass electron-positron states. Each of these cases involves not only selectivity, but also discordant experimental results and it was demonstrated that at least some of the experimental results were artifacts caused by selectivity. I present detailed histories showing how the artifactual nature of the results was established. For each of these three episodes, I also discuss how the discord between the experimental results was resolved. This provides an introduction to Part II, which deals with the resolution of discordant results.

Part I also includes a discussion of blind analysis, a strategy that is in current use and is designed to eliminate the possibility of experimenter bias. I examine how one argues that the use of Monte Carlo calculations, an analysis procedure often used in the production of experimental results, does not produce artifacts. In such calculations, physicists often have choice of both input parameters and analysis techniques, raising the possibility of bias. I examine the methods used to guard against this possibility.

As noted above, it is a fact of life in empirical science that experiments often give discordant results. The occurrence of such discordant results casts doubt on my epistemology of experiment and on the reasonable use of experimental results in science. If, as is the case, each of the experiments involved applied the epistemology of experiment, how can they produce discordant results? In Part II, I examine several additional episodes from contemporary science that include discordant results. I discuss the strategies and arguments used to resolve that discord and show that they do not cast doubt on the reasonable use of experimental results in science.

I

SELECTIVITY AND THE PRODUCTION OF EXPERIMENTAL RESULTS

Any fool can take data. It's taking good data that counts.

—E. Commins (private communication)

Experimenters never use all of their data in producing a result. Data may be excluded for many legitimate reasons.[1] Certainly no one would think of using data obtained when the experimental apparatus is not working properly. Even when the apparatus is working properly, problems may arise when only selected portions of the data (i.e., "good" data) are used to obtain a result. Selection criteria, usually referred to as "cuts," are applied to either the data themselves or the analysis procedures[2] and are designed to maximize the desired signal and to eliminate or minimize background that might mask or mimic the desired effect. A legitimate concern is that the experimental result may be an artifact produced by the cuts and thus not a valid result.[3] A further cause for concern may arise if the effect of the cuts on the experimental result is known in advance: Is the experimenter tuning the cuts to produce a desired outcome?

This is not a purely philosophical or methodological issue: It is basic to the actual practice of science. It was the central issue in the recent controversy concerning the possible existence of low-mass electron-positron states, or particles, produced in high-energy heavy ion-atom collisions (see Chapter 5). Because selection cuts are invariably present in modern physics experiments, the question of how one argues that an observed effect is not an artifact produced by the cuts is often of crucial importance in establishing the validity of experimental results. Experimenters use several strategies to argue that their result is not an artifact. These strategies are illustrated in Chapters 1–5, which illuminate the different ways in which selectivity is used in producing experimental results and discuss the arguments for the validity of the application of the cuts.

The first of these strategies involves robustness. Experimenters vary the selection criteria over reasonable limits and observe whether the result is stable under such variations. If it is, then the result is taken to be real. In an experiment to measure the K_{e2}^{+} branching ratio, the fraction of all K^{+} mesons that decay into a positron plus a neutrino, the experimenters varied the values of a range cut, as well as their track matching criteria, and found that the branching ratio remained constant (see Chapter 1). Similarly, in the case of early attempts to detect gravity waves, one of the selection criteria involved a choice of analysis algorithm. One group of experimenters used both proposed algorithms and found no change in their result (see Chapter 2). In the case of the 17-keV neutrino, there was a choice to be made concerning the energy range to be used in the analysis of the data. Several experiments used both a wide- and a narrow-energy range and showed the result to be constant (discussed in Chapter 4).

If the result is sensitive to variations in the selection criteria, then this suggests—although it does not prove—that the result is an artifact. There are indeed cases, such as resonant scattering of light, in which the result is highly sensitive to the experimental conditions and also to the particular selection criteria. This type of sensitivity will be an important issue in the episode of possible low-mass $e^{+}e^{-}$ states discussed in Chapter 5, in which the observed effects seemed to be very sensitive to the experimental conditions.

Robustness may also be provided by a sequence of experiments, rather than by variations on a single experiment. Thus in the case of Millikan's measurement of the charge on the electron (Chapter 3), the variations of the selection criteria he applied to both his data and to his analysis procedures were provided by subsequent experiments. These experiments used different experimental techniques and had different backgrounds and selection criteria. That Millikan's experiment and the subsequent experiments all gave the same value for the charge of the electron argued that the result was not an artifact produced by the cuts. If the result had been an artifact, it is highly improbable that the same result would be obtained under such very different circumstances. The replicability, or apparent replicability, of results also played an important role in the episode concerning possible low-mass e^+e^- states.

Sometimes it is possible to use a surrogate signal to demonstrate that the cuts do not mask the presence of an effect.[4] This was the case in both the search for gravity waves and for a 17-keV neutrino. Conversely, one may cast doubt on a result by showing that it can be spuriously produced by the application of cuts (as in the case of the low-mass electron-positron states, discussed in Chapter 5). This may be done either by the analysis of actual data or by computer simulation. This method of questioning the validity of a result is illustrated several times in the following chapters, and is crucial in the discussion of low-mass electron-positron states.

One may also be able to argue that the application of cuts, although reducing background, cannot produce the effect observed. In the case of the measurement of the K_{e2}^+ branching ratio, there was no possible way in which the application of cuts to the range, decay time, or track-matching of the particles could produce positrons that would mimic those expected from K_{e2}^+ decay.

In Part I, I discuss five historical cases that involve arguments concerning the reality of an experimentally observed effect. To introduce the reader to the issues, I begin with four straightforward examples taken from the history of modern physics. The first is an experiment designed to measure the K_{e2}^+ branching ratio (Chapter 1). The branching ratio is quite small and the desired events were masked by large numbers of events due to other, more common decay modes. Cuts were applied to preferentially reduce this background while preserving a large, and known, fraction of the K_{e2}^+ events. I show how the experimenters used the cuts to produce the experimental result and how they argued for its validity.

A somewhat more complex example is provided by Joseph Weber's claim that he had observed gravity waves, whereas six other experiments did not find his claimed effect (Chapter 2). In this episode, there were no arguments about what constituted good data. Both Weber and his critics used the same type of experimental apparatus, namely, a large-mass gravity-wave antenna known as a Weber bar. The question was whether the data were being analyzed correctly. Weber and his critics used different data-analysis algorithms. The linear algorithm used by Weber's critics was sensitive to changes in either the amplitude or the phase of the signal. The nonlinear algorithm preferred by Weber was sensitive only to changes in amplitude. There was a suggestion that Weber chose his algorithm because it gave a larger gravity-wave signal in his experiment. A second question concerned the pulse-height threshold that was used to determine whether a signal was in fact present. Weber's use of varying thresholds raised the issue of whether he was tuning his threshold cut to maximize—or even to create—evidence for his positive signal. I discuss how these issues were decided and the discord resolved. This episode will also emphasize the importance of paying close attention to analysis procedures in order to understand arguments concerning the validity of an experimental result.

The third historical episode concerns Millikan's famed measurement of e, the charge of the electron (Chapter 3). Examination of Millikan's notebooks has shown not only that Millikan excluded data, but that he also engaged in selective calculational procedures. We also know that Millikan had a clear expectation of the value of e, based on his earlier work. One might ask whether he used selectivity to obtain the answer he expected. I discuss the effect of what we might call Millikan's "cosmetic surgery" on his experimental result.

In the fourth case, I look at the role played by analysis cuts in the episode that decided against the existence of the 17-keV neutrino, a proposed new particle (Chapter 4). The last and longest of the historical cases examines in detail the history of low-mass electron-positron states, from their initial report in the early 1980s to the present (Chapter 5).

Part I concludes with a chapter on blind analysis, a technique used to guard against experimenter bias, which includes a discussion of Monte Carlo simulations (Chapter 6). As we shall see, these calculations are often crucial in producing experimental results. Some critics have questioned the use of such calculations in the production of experimental results, because the experimenters can choose the parameters used in the calculations. I discuss how one establishes the correctness of such calculations.

1

Measurement of the K^+_{e2} Branching Ratio

Perhaps the simplest and most straightforward strategy used to argue for the correctness of a result when selection criteria are used in its production is to vary the values of the cuts being used. If the result remains constant under such variation then it can be argued that the outcome is not an artifact of the cuts: if the effect is real, then reasonable changes in the cuts used should not affect the result. This strategy is clearly illustrated in an experiment designed to measure the K^+_{e2} branching ratio, defined as the fraction of all K^+ mesons that decay into a positron and an electron neutrino ($K^+ \rightarrow e^+ + \nu_e$) (Bowen et al., 1967).

The motivation for this experiment was that it would be a stringent test, using strangeness-changing decays, of the then-generally accepted V–A (vector minus axial vector) theory of weak interactions. At the time of this experiment, the V–A theory had strong experimental support, although it had not been severely tested in strangeness-changing decays.[1]

The theoretical predictions of the K^+_{e2} branching ratio were explicit. If the interaction was pure axial vector, the predicted ratio of K^+_{e2} to $K^+_{\mu 2}$ decays was 2.6×10^{-5}, corresponding to a branching ratio of 1.6×10^{-5}. Pure pseudoscalar coupling, however, predicted a K^+_{e2} to $K^+_{\mu 2}$ ratio of 1.02. Thus even if only a small portion of the interaction was due to

---

For a more detailed discussion of this experiment, see Franklin (1990, pp. 118–31).

pseudoscalar coupling, the K^+_{e2} branching ratio would be much larger than that for a pure axial-vector interaction. For example, adding only one part in a thousand of pseudoscalar interaction to the axial-vector interaction would increase the expected branching ratio by a factor of four. Thus, even a rough measurement of the K^+_{e2} branching ratio would be a stringent test for the presence of any pseudoscalar interaction in the decay, and of the V–A theory in general. The best previous measurement of the $K^+_{e2}/K^+_{\mu2}$ ratio had set an upper limit of 2.6×10^{-3}, a factor of 100 larger than that predicted by V–A theory.

In principle, this is a simple experiment. The positron from K^+_{e2} decay has a momentum of 246.9 MeV/c in the kaon center-of-mass frame of reference. This value is larger than the momentum of any other charged particle produced in the direct decay of the kaon, the next largest being the muon from $K^+_{\mu2}$ decay (235.6 MeV/c). Thus all one has to do, in principle, to measure the branching ratio is to stop the K meson, identify the decay particle as a positron, and measure its momentum. If the momentum is approximately 247 MeV/c, then the positron is the product of K^+_{e2} decay, and the event can be added to the count of K^+_{e2} decays. One would then compare this tally to the total number of kaon decays to obtain the branching ratio. In practice, however, the experiment was far more difficult.

Experimental Apparatus

The experimental apparatus is shown in Figure 1.1. The incoming beam was positively charged, unseparated, and momentum selected. It consisted primarily of pions and protons, and also contained small numbers of kaons, muons, and positrons. The desired kaons were separated from the more numerous pions and protons by range in matter and by time of flight[2] and were stopped in counter C_3, the stopping region. (For details, see Bowen et al., 1967.)

Decay particles that left the stopping region at about 90° to the incoming beam traversed a set of six thin-plate optical spark chambers located in a magnetic field. This allowed a measurement of the particle's momentum. Decay particles were detected by coincidence telescope C_5C_6. Pulses from counters C_3 and C_5 were displayed as oscilloscope traces and photographed, so that the time between them, which measured the time interval between the K^+ stop and its decay, could be measured for

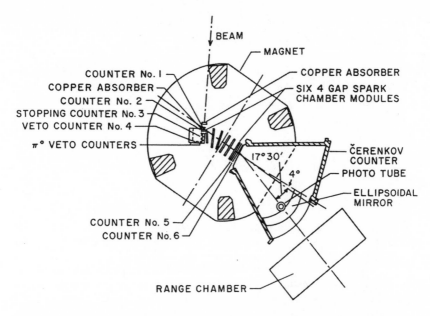

Figure 1.1. Experimental apparatus for the measurement of the K$^+_{e2}$ branching ratio. From Bowen et al. (1967).

each event. If the decay particles were indeed due to kaon decay, then the distribution of time intervals between the stopping kaon signal and the decay signal should match the kaon lifetime. It did. A small prompt peak was observed at short decay times due to kaon decays in flight. These were completely eliminated during the analysis procedure by a requirement that the decay time of the event be >2.75 ns after the stop of the K$^+$ meson.

This apparatus incorporated both a Čerenkov counter, to identify positrons, and a range chamber to help eliminate background from other decay modes. The Čerenkov counter had a measured efficiency of >99% for positrons and of 0.38% for other particles of comparable momentum. The thick-plate range chamber was placed behind the Čerenkov counter, which permitted measurement of the position of particles emerging from the counter as well as a measurement of their total range.

Selection Criteria

There were three major sources of background events that might mimic or mask the desired K$^+_{e2}$ events. These were:

1. Accidental coincidences between accelerator-produced background in the Čerenkov counter and muons and from $K^+_{\mu 2}$.

2. $K^+ \rightarrow \mu^+ + \nu_\mu$, followed by $\mu^+ \rightarrow e^+ + \nu_e + \nu_\mu$, with a maximum momentum of 246.9 MeV/c (the same as that for K^+_{e2} decay) and a branching ratio of approximately 1.2×10^{-4} per foot of muon path. This decay rate per foot was considerably larger than the total expected K^+_{e2} decay rate. If this source of background could not be eliminated or considerably reduced, then the experiment could not be done.

3. Decays during the flight of the K^+ meson.

Figure 1.2 shows the momentum distribution of 16,965 events obtained with the Čerenkov counter in the triggering logic.[3] This is the haystack from which the needle of a few K^+_{e2} events was to be found. (A rough calculation indicated that approximately five K^+_{e2} events would be observed.) The momentum for K^+_{e2} decay is shown. It is clear that if the K^+_{e2} events are present they are rather well hidden.

The large peak at 236 MeV/c and the smaller peak at 205 MeV/c are due to accidental coincidences between accelerator-produced background in the Čerenkov counter and muons and pions from $K^+_{\mu 2}$ and $K^+_{\pi 2}$ decay, respectively. If the K^+_{e2} events were to be found, then the background due to $K^+_{\mu 2}$ events had to be reduced.

The experimenters applied a set of criteria to eliminate unwanted background events while preserving a reasonable and known fraction of the K^+_{e2} events. The first criterion applied was that of range—the path length in the range chamber before the particle stopped or underwent an interaction. This criterion was designed to reduce the number of events from $K^+_{\mu 2}$ decay. Muons lose energy only by ionization loss and thus have a well-defined range in matter. The muons from $K^+_{\mu 2}$ had a mean measured range of 67 g/cm^2, with a straggle of about 4 g/cm^2 (Figure 1.3). The experimenters measured the range distribution for positrons with momenta between 212 MeV/c and 227 MeV/c (Figure 1.4). These positrons differ from K^+_{e2} positrons by only 10% in momentum and were expected to behave quite similarly. Positrons do not have a well-defined range because they lose energy by several different processes, some of which involve large energy losses, and the distribution of ranges is approximately constant from about 15 g/cm^2 to 70 g/cm^2. Requiring events to have a range less than that of the muon from $K^+_{\mu 2}$ decay serves

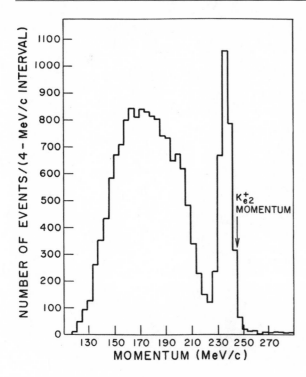

Figure 1.2. Momentum distribution for all K^+ decay events obtained with the Čerenkov counter in the triggering logic. From Bowen et al. (1967).

to minimize the background due to those events and yet preserves a large (and known) fraction of the high-energy positrons. A selection cut was made at 45 g/cm^2. The limits on this cut were varied within reasonable limits (±5 g/cm^2); it was found that the final result was robust against such changes. The effect of applying this selection criterion to the data is shown in Figure 1.5. The haystack has gotten smaller.

Another major source of background was decay of the kaon into a muon, followed by the decay of the muon into a positron. Most of these positrons are emitted at large angles to the muon path. If the decay occurred in the momentum chambers, it would have been detected by a kink in the track (see discussion in note 3). Decays occurring between the end of the momentum chambers and the end of the Čerenkov counter— a very long distance—could not be seen. Because of the usually large decay angle, such decays could be detected by comparing the measured position of the particle when it entered the range chamber with the position predicted by extrapolating the momentum-chamber track. If a decay had occurred, then the difference between the two positions would be

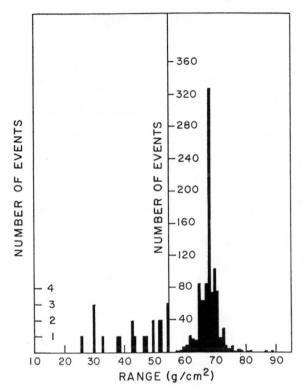

Figure 1.3. Range spectrum for muons from $K^+_{\mu 2}$ decay. From Bowen et al. (1967).

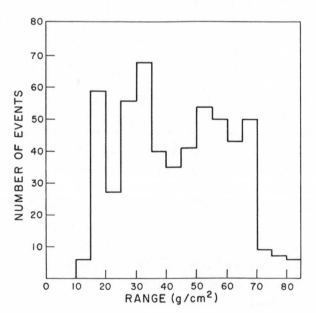

Figure 1.4. Range spectrum for positrons from K^+_{e3} decay with momentum between 212 and 227 MeV/c. From Bowen et al. (1967).

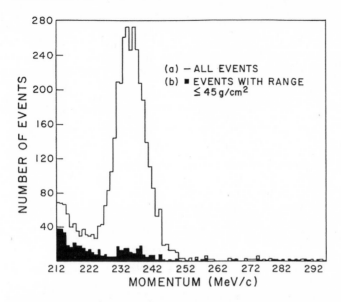

Figure 1.5. Momentum spectrum of particles with momentum >212 MeV/c: (a) all events; (b) events with range ≤45 g/cm². From Bowen et al. (1967).

large. Even for decay angles as small as 5°, the extrapolated momentum-chamber track will not match the position of the range-chamber track. Therefore, a cut on the difference between the measured and extrapolated positions, D_x and D_y (defined as the x and y differences, respectively) can be used to eliminate muon decays in flight. If decays with decay angles >5° are eliminated, then the background due to muon decays in flight would be reduced to approximately 5% of the expected K^+_{e2} rate. The experimental distributions for D_x and D_y are shown in Figures 1.6 and 1.7 for positrons resulting from K^+_{e3} decay ($K^+ \rightarrow e^+ + \nu_e + \pi^0$; these positrons do not decay). The width of these distributions and the accuracy of the comparison were limited by multiple scattering in the momentum chambers and by the uncertainty in extrapolating the particle trajectory through the fringing field of the magnet. The full width at half maximum of the distributions is 16 cm for D_x and 13 cm for D_y. Fiducial areas −6 cm to +10 cm for D_x and −7 to +6 cm for D_y were chosen, as indicated in Figures 1.6 and 1.7. Table 1.1 shows the variation in the number of accepted particles in the momentum regions of interest as the fiducial areas were varied. The ratios are constant within the calculated statistical uncertainty, showing that the branching ratio was robust under

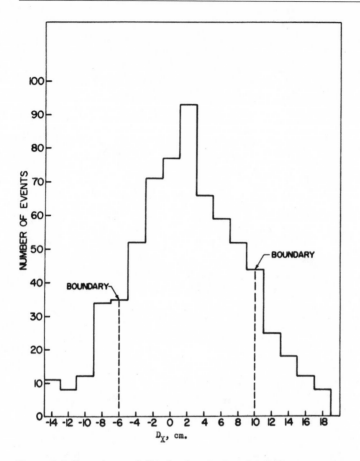

Figure 1.6. Experimental distribution for D_x, the difference in x-position between the track extrapolated from the momentum chambers and the measured track in the range chamber, for positrons from K^+_{e3} decay with momenta between 207 and 227 MeV/c. These positrons do not decay in flight. From Bowen et al. (1967).

these changes in the cuts. On the basis of these results, the track-matching criterion was applied. In addition, because these decays occurred beyond the momentum chambers, they would have a measured momentum equal to that of muons from $K^+_{\mu 2}$ decay (236 MeV/c). Some of these events would be removed from the data by momentum cuts in the final analysis of the data. The effects of this track-matching cut for events with range ≤ 45 g/cm^2 and momentum ≥ 212 MeV/c are shown in Figure 1.8. The selection criteria served to preferentially reduce the events in the $K^+_{\mu 2}$ region relative to the events in the K^+_{e2} region. (These regions are

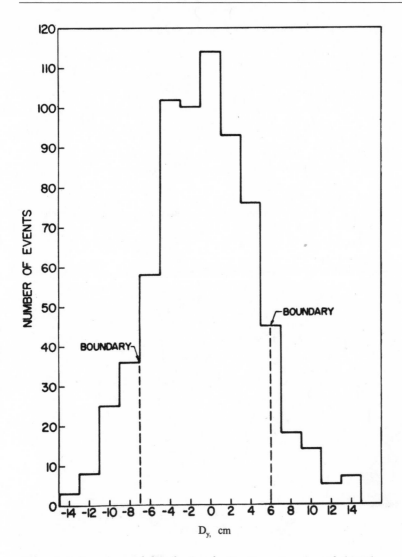

Figure 1.7. Experimental distribution for D_y. From Bowen et al. (1967).

232–242 MeV/c and 242–252 MeV/c, respectively. They are centered on the momenta for the respective decays.)

There is one additional major source of background. This is due to decays in flight of the K^+ meson. If the kaon decayed in flight, then the momentum of the decay particle could be increased, leading to possible simulation of K_{e2}^+ decays. Examination of the distribution of time inter-

Table 1.1.
Effect of the track-matching criteria for events with range <45 g/cm² [a]

Momentum Region (MeV/c)	None	Fiducial Area		
		Accepted Intervals	Accepted Intervals Increased by 2 cm	Accepted Intervals Decreased by 2 cm
$P \leqslant 212$	525	214	240	184
K^+_{e3} (212–228)	297	161	177	143
$K^+_{\mu2}$ (231–241)	134	28	35	20
K^+_{e2} (242–252)	33	13	13	11
$P \geqslant 252$	36	5	6	4

Source: Bowen et al. (1967).
[a]The number of events in the momentum regions of interest does not change significantly for small variations of the fiducial area.

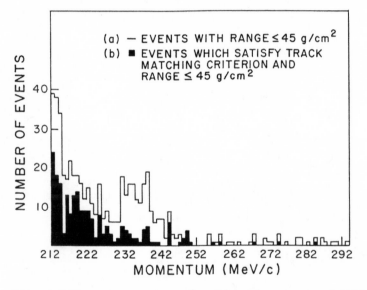

Figure 1.8. Momentum spectrum of particles with momentum $\geqslant$212 MeV/c and range $\leqslant$45 g/cm²: (a) all events; (b) events satisfying track-matching criterion. From Bowen et al. (1967).

vals between the stopped kaon and the decay positron revealed the presence of a small peak due to such decays in flight. The peak had a base width of 2 ns. A cut was made removing all events with a time interval of <2.75 ns, which eliminated all of the decays in flight. The effect of this selection criterion is shown in Figure 1.9. This cut preferentially reduced

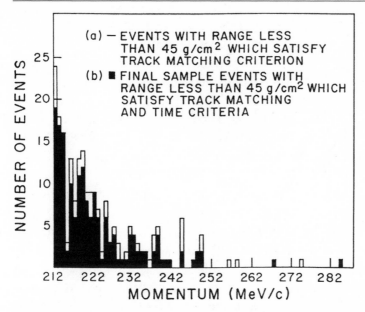

Figure 1.9. Momentum spectrum of particles with momentum ≥212 MeV/c, range ≤45 g/cm², and satisfying the track-matching criterion: (a) all events; (b) events with K⁺ decay time ≥2.75 ns. From Bowen et al. (1967).

the number of events in the K^+_{e2} region, indicating that decays in flight were indeed a source of simulated K^+_{e2} events. The cut was not varied because it was intended solely to eliminate decays in flight. It was clear from the decay-time distribution, which showed a small prompt peak that was 2 ns wide, that the cut at 2.75 ns was sufficient. In addition, it affected both the number of K^+_{e2} events and the events used to normalize the branching ratio equally; hence, changing the cut would not affect the branching ratio. The final number of K^+_{e2} candidates measured by the experiment tallies the events in the momentum region 242–252 MeV/c in Figure 1.9.

The number of events in the K^+_{e2} region is corrected for various experimental effects to determine the final number of K^+_{e2} events. A final total of $6^{+5.2}_{-3.7}$ events is attributed to K^+_{e2} decay after these corrections. The branching ratio—the rate compared with all K⁺ decays—was calculated by normalizing the K^+_{e2} events to known K⁺ decay rates by two different methods. The first used the upper end of the K^+_{e3} spectrum (the region from 212 MeV/c to 228 MeV/c in Figure 1.9), which had been subjected to the same selection criteria as the K^+_{e2} events. To estimate the total number of K^+_{e3} events, the experimenters needed to know the shape of the K^+_{e3} decay spectrum. This had, in fact, been measured by the group in previous experiments. The second method used the total sample of 16,965 K^+ decays given in Figure 1.2. (Note that the selection criteria have not been applied to these events.) The results for the branching ratio, using the two different methods, were $R = 2.0^{+1.8}_{-1.2} \times 10^{-5}$ and $R = 2.2^{+1.9}_{-1.4} \times 10^{-5}$, respectively. The two different methods—which have very different selection criteria—agreed and the final result given was their average, $R = 2.1^{+1.8}_{-1.3} \times 10^{-5}$, in agreement with the theoretical prediction of 1.6×10^{-5}.

Thus the branching ratio obtained was robust under these two very different normalization methods, which had very different dependences on the selection criteria. It was also robust under reasonable changes in the selection criteria. These observations suggest that the cuts did not affect the final result and therefore they argue for the correctness of that result. The experimental results are a typical and straightforward example of robustness.

2

Early Attempts to Detect
Gravity Waves

In the previous chapter, I discussed an episode in which selection criteria were applied to experimental data. The present chapter deals with a case in which selectivity was applied to the analysis procedures used to transform data into an experimental result. No questions were raised as to what constituted good data. All of the experimental groups used similar types of experimental apparatus and agreed that they produced good data. Here we will see how different analysis procedures led to discordant results, and I discuss how that discord was resolved. The episode considered here comprises the early attempts to detect gravity waves.

Beginning in the late 1960s, attempts were made to detect gravitational radiation (gravity waves). Such waves are predicted by Einstein's general theory of relativity. Just as an accelerated, electrically charged particle will produce electromagnetic radiation (light, radio waves, etc.), so should an accelerated mass produce gravitational radiation. Such radiation can be detected by the oscillations produced in a large mass when it is struck by gravity waves. Because the gravitational force is far weaker than the electromagnetic force, a large mass must be accelerated to produce a detectable gravity-wave signal.[1] The difficulty of detecting a weak signal is at the heart of this episode.

For more details of this episode, see Franklin (1994). For a very different view, see Collins (1985, 1994).

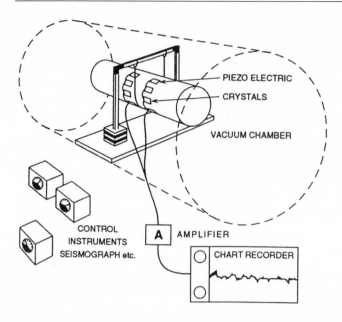

Figure 2.1. A Weber-type gravity-wave detector. From Collins (1985).

In 1969, Joseph Weber claimed to have detected such radiation. Weber used a massive aluminum-alloy bar,[2] or antenna, which was supposed to oscillate when struck by gravitational radiation (Figure 2.1). The oscillation was to be detected by observing the amplified signal from piezoelectric crystals attached to the antenna. The signals were expected to be quite small, and the bar had to be insulated from other sources of noise, such as electrical, magnetic, thermal, acoustic, and seismic forces. Because the bar was at a temperature different from absolute zero, thermal noise could not be avoided; to minimize its effect, Weber set a threshold for pulse acceptance. Weber claimed to have observed above-threshold pulses (in excess of those that are to be expected above the threshold from thermal noise).[3] In 1969, Weber claimed to have detected approximately 7 pulses/day due to gravitational radiation.

The problem was that Weber's reported rate was far greater than that expected from calculations of cosmic events (by a factor of more than 1,000), and his early claims were met with skepticism. During the late 1960s and early 1970s, however, Weber introduced several modifications and improvements that increased the credibility of his results (Weber et al., 1973). He claimed that above-threshold peaks had been observed simultaneously in two detectors separated by 1,000 miles. Such coincidences were extremely unlikely if they were due to random thermal fluctuations. In addition, he reported a 24-hour periodicity in his peaks—the sidereal correlation—that indicated a single source for the radiation, perhaps near the center of our galaxy. These results increased the plausibility of his claims sufficiently so that by 1972, three other experimental groups had not only built detectors, but had also reported results. None was in agreement with Weber. By 1975, it was generally agreed that Weber's claim was unacceptable.

The reasons offered by different scientists for their rejection of Weber's claims are varied, and not all of the scientists engaged in the pursuit agreed about their relative significance. During the period 1972–1975, it was discovered that Weber had made several serious errors in his analysis. His computer program for analyzing the data contained an error, and his statistical analysis of residual peaks and background was questioned and thought to be inadequate. Weber's claim to have found coincidences between his detector and a second, distant detector was rejected because the tapes used to provide the coincidences were actually recorded more than four hours apart. Weber had found a positive result where no one would expect one. Other critics cited the failure of Weber's signal-to-noise ratio to improve, despite the "improvements" to his apparatus. In addition, the sidereal correlation previously observed disappeared when more data were taken, suggesting that it was a statistical fluctuation in the data.[4] Critics also argued that Weber's apparatus, as described in his published work, could not produce the signal he reported. Perhaps the most important objection was the uniformly negative results obtained by six other groups of experimenters.[5]

Two of the reasons for the rejection of Weber's result by the physics community involve questions concerning selection criteria.[6] The first of these—the issue of calibration together with Weber's analysis procedure—is not so much a selection criterion applied to the data as a choice of the analysis procedure used. The problem of determining whether there is a signal in a gravity-wave detector, or whether two such detectors have fired simultaneously is not simple. There are several difficulties. One is due to energy fluctuations in the bar from nongravitational sources (e.g., thermal, acoustic, electrical, magnetic, and seismic noise). When a gravity wave strikes the antenna, the wave's energy is added to that already present in the bar. This may change either the amplitude or the

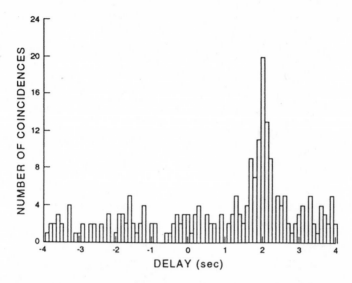

Figure 2.2. A plot showing the calibration pulses for the Rochester-Bell Laboratory collaboration. The peak due to the calibration pulses is clearly seen. From Shaviv and Rosen (1975).

phase (or both) of the signal emerging from the bar. It is not simply a case of observing a larger signal from the antenna after a gravity wave strikes it. This difficulty informs the discussion of which was the best analysis procedure to use.

The nonlinear, or energy, algorithm preferred by Weber is sensitive only to changes in the amplitude of the signal. The linear algorithm, preferred by everyone else, is sensitive to changes in both the amplitude and the phase of the signal. Weber admitted, however, that the linear algorithm preferred by his critics is more efficient (by a factor of twenty) at detecting calibration pulses. These were pulses of acoustic energy injected into the antenna to simulate the effect of gravity waves and to test whether the apparatus was working properly. Similar results on the superiority of the linear algorithm for detecting calibration pulses were reported by both Kafka (pp. 258–59) and Tyson (pp. 281–82). Tyson's results for calibration-pulse detection are shown for the linear algorithm in Figure 2.2 and for the nonlinear algorithm in Figure 2.3. There is a clear peak for the linear algorithm, whereas no such peak is apparent for the nonlinear procedure. (The calibration pulses were inserted periodically during data-taking runs. The peak was displaced by two seconds by

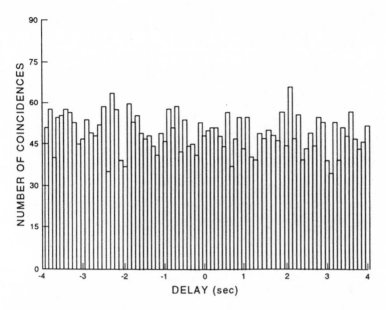

Figure 2.3. A time-delay plot for the Rochester-Bell Laboratory collaboration, using the nonlinear algorithm. No sign of any zero-delay peak is seen. From Shaviv and Rosen (1975).

the insertion of a time delay, so that the calibration pulses would not mask any possible real signal, which was expected at zero time delay.)

Nevertheless, Weber preferred the nonlinear algorithm. His reason for this was that it gives a more significant signal than does the linear procedure. This is illustrated in Figure 2.4, which compares the data analyzed using the nonlinear and the linear algorithms. Weber (pp. 251–52) remarked, "Clearly these results are inconsistent with the generally accepted idea that $\dot{x}^2 + \dot{y}^2$ [the linear algorithm] should be the better algorithm." Weber was, in fact, using the positive result to decide which was the better analysis procedure. He was tuning his analysis procedure to maximize his result.

His critics, however, analyzed their own data using both algorithms. If it was the case that—unlike the calibration pulses, for which the linear algorithm was superior—using the linear algorithm either masked or failed to detect a real signal, then using the nonlinear algorithm on their data should produce a clear signal. None appeared. Typical results are shown in Figures 2.3 and 2.5. Figure 2.3, which is Tyson's data analyzed with the nonlinear algorithm, not only shows no calibration peak, but it

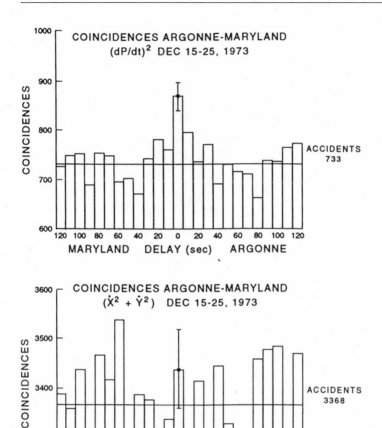

Figure 2.4. Weber's time-delay data for the Maryland-Argonne collaboration for the period 15–25 December 1973. The top graph uses the nonlinear algorithm, whereas the bottom uses the linear algorithm. The zero-delay peak is seen only with the nonlinear algorithm. From Shaviv and Rosen (1975).

does not show a signal peak at zero time delay. It is quite similar to the data analyzed with the linear algorithm shown in Figure 2.5. (Note that for this data run, no calibration pulses were inserted.) Kafka (pp. 258–59) also reported the same result: no difference in signal between the linear and the nonlinear analysis.

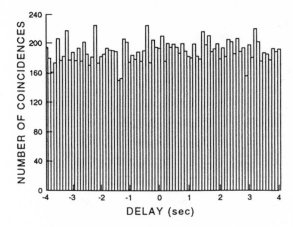

Figure 2.5. A time-delay plot for the Rochester-Bell Laboratory collaboration, using the linear algorithm. No sign of a zero-delay peak is seen. From Shaviv and Rosen (1975).

Weber answered these criticisms by suggesting that although the linear algorithm was better for detecting calibration pulses, which were short, the real signal of gravity waves was a longer pulse than most investigators thought. He argued that the nonlinear algorithm that he used was better at detecting these longer pulses. Still, if the signal was longer, one would have expected it to show up when the critics' data were processed with the nonlinear algorithm. It did not. (See Figures 2.3 and 2.5.) The critics' results were robust under changes in the analysis procedure.

Drever also reported that he had looked at the sensitivity of his apparatus with arbitrary waveforms and pulse lengths. Although he found a reduced sensitivity for longer pulses, he did analyze his data explicitly to look for such pulses. He found no effect with either the linear or nonlinear analysis.[7]

How, then, did Weber obtain his positive result when his critics, using his own analysis program, could not? It was suggested that Weber had varied his threshold cut to maximize his signal, whereas his critics used a constant threshold. Was Weber tuning his threshold cut to create a result? This was the second reason why critics rejected Weber's result.

Tyson characterized the difference between Weber's methods and those of his critics:

I should point out that there is a very important difference in essence in the way in which many of us approach this subject and the way Weber approaches it. We have taken the attitude that, since these are integrating calorimeter type experiments which are not too sensitive to the nature of pulses put in, we simply maximize the sensitivity and use the algorithms which we found maximized the signal to noise ratio, as I showed you. Whereas Weber's approach is, he says, as follows. He really does not know what is happening, and *therefore he or his programmer is twisting all the adjustments in the experiment more or less continuously, at every instant in time locally maximizing the excess at zero time delay.* I want to point out that there is a potentially serious possibility for error in this approach. No longer can you just speak about Poisson statistics. *You are biasing yourself to zero time delay, by continuously modifying the experiment on as short a time scale as possible (about four days), to maximize the number of events detected at zero time delay.* We are taking the opposite approach, which is to calibrate the antennas with all possible known sources of excitation, see what the result is, and maximize our probability of detection. Then we go through all of the data with that one algorithm and integrate all of them. Weber made the following comment before and I quote out of context: "Results pile up." I agree with Joe (Weber). But I think you have to analyze all of the data with one well-understood algorithm. (p. 293, emphasis added)

A similar criticism was offered by Garwin, who also presented evidence from a computer simulation to demonstrate that a selection procedure such as Weber's could indeed produce his positive result:

Second, in view of the fact that Weber at CCR-5 [a conference on General Relativity held in Cambridge][8] explained that when the Maryland group failed to find a positive coincidence excess "we try harder," *and since in any case there has clearly been selection by the Maryland group* (with the publication of data showing positive coincidence excesses but with no publication of data that does not show such excesses),[9] James L. Levine has considered an extreme example of such selections. In Figure [2.6] is shown the combined histogram of "coincidences" between two independent streams of random computer-generated data. This "delay histogram" was obtained by partitioning the data into 40 segments. For each segment, "single events" were defined in each "channel" by assuming one of three thresholds a, b, or c. That combination of thresholds was chosen for each segment which gave the maximum "zero delay coincidence" rate for that segment. The result was 40 segments selected from one of nine "experiments." *The 40 segments are summarized in Figure [2.6], which shows a "six-standard-deviation" zero-delay excess.* (Garwin, 1974, pp. 9–10, emphasis added)

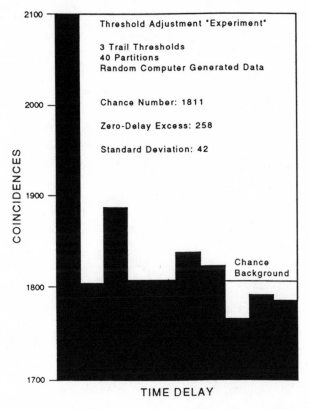

Figure 2.6. The result of selecting thresholds that maximized the zero-delay signal for Levine's computer simulation. From Garwin (1974).

Weber denied the charges:

It is not true that we turn our knobs continuously. I have been full time at the University of California at Irvine for the last six months, and have not been turning the knobs by remote control from California [Weber's group and one of his antennas was located at the University of Maryland]. In fact, the parameters have not been changed for almost a year. What we do is write the two algorithms on a tape continuously. The computer varies the thresholds to get a computer printout which is for 31 different thresholds. The data shown are not the results of looking over a lot of possibilities and selecting the most attractive ones. We obtain a result that is more than three standard deviations for an extended period for a wide range of thresholds. I think it is very important to take the point of view that the histogram itself is the final judge of what the sensitivity is. (pp. 293–94)

Weber did not, however, specify his method of data selection for his histogram. In particular, he did not state that all of the results presented in a particular histogram had the same threshold.

Interestingly, Weber cited evidence provided by Kafka as supporting a positive gravity-wave result. Kafka did not agree, because the evidence resulted from performing an analysis using different data segments and different thresholds. Only one data segment/threshold showed a positive result, indicating, in fact, that such selectivity could produce a positive result. Kafka's results are shown in Figure 2.7. Note that the positive effect is seen in only the bottom graph:

The very last picture (Figure [2.7]) is the one in which Joe Weber thinks we have discovered something, too. This is for 16 days out of 150. There is a 3.6σ [standard deviation] peak at zero time delay, but you must not be too impressed by that. It is one out of 13 pieces for which the evaluation was done, and I looked at least at 7 pairs of thresholds. Taking into account selection we can estimate the probability to find such a peak accidentally to be of the order of 1%. (p. 265)

In this episode, it was suggested that Weber's positive result was caused, in part, by tuning both his analysis procedure and his threshold cut to produce that result. Weber's critics dealt with this problem by analyzing their data using both their own linear analysis algorithm and Weber's preferred nonlinear algorithm; they obtained negative results with both procedures. The results were robust. The critics also showed how one might produce a positive result by tuning the threshold cut. This was shown by Kafka, using his actual data, and by Levine and Garwin, who obtained a positive result from random data in a Monte Carlo simulation[10] by manipulating the threshold cut. The latter arguments did not conclusively demonstrate that Weber's results were incorrect, but they strongly suggested that there were credibility problems with his results. Weber's selectivity had led to discordant results. How was the discord between Weber and his critics resolved?

Let us summarize the evidential situation concerning gravity waves at the beginning of 1975. There were discordant results. Weber had reported positive results on gravitational radiation, whereas six other groups had reported no evidence for such radiation. The critics' results were not only more numerous, but had also been carefully cross-checked. The groups had exchanged both data and analysis programs and confirmed the results. The critics had also investigated whether their analysis

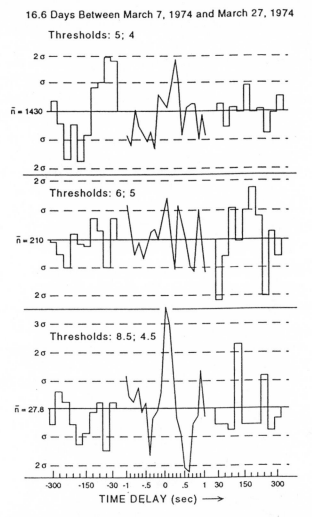

16.6 Days Between March 7, 1974 and March 27, 1974

Figure 2.7. Kafka's results using various thresholds. A clear peak is seen at zero delay. From Shaviv and Rosen (1975).

procedure (the use of a linear algorithm) could account for their failure to observe Weber's reported results. They had used Weber's preferred procedure (a nonlinear algorithm) to analyze their data and still found no sign of an effect. They had also calibrated their experimental apparatuses by inserting acoustic pulses of known energy and finding that they could detect a signal. Weber, however, could not detect such calibration pulses (neither could his critics when they used his analysis procedure). Under ordinary circumstances, Weber's calibration failure would have been decisive. Because this episode is atypical—one in which a new type of apparatus was used to search for a hitherto unobserved phenomenon—the calibration failure was not decisive. Other arguments were both needed and provided to resolve the discord.

The physics community raised several other serious questions about Weber's analysis procedures. The various experimental groups cooperated, exchanging both data and analysis programs. This led to the first of several questions concerning possible serious errors in Weber's analysis of his data. Douglass et al. (1975) pointed out that Weber's analysis program contained an error. It generated coincidences between detectors even when none were present. Douglass also pointed out that this error accounted for all of the coincidences observed in the tape of Weber's data that he had examined. Weber admitted the error, but did not agree with the claim that it accounted for all of his observed coincidences. At the very least, this error raised legitimate doubts about Weber's results.

There was also a rather odd result reported by Weber, discussed here by Garwin (1974):

First, Weber has revealed at international meetings (Warsaw, 1973. etc.) that he had detected a 2.6-standard deviation excess in coincidence rate between a Maryland antenna [Weber's apparatus] and the antenna of David Douglass at the University of Rochester. Coincidence excess was located not at zero time delay but at "1.2 seconds," corresponding to a 1-sec intentional offset in the Rochester clock and a 150-millisecond clock error. At CCR-5, Douglass revealed, and Weber agreed, that the Maryland Group had mistakenly assumed that the two antennas used the same time reference, whereas one was on Eastern Daylight Time and the other on Greenwich Mean Time. Therefore, the "significant" 2.6 standard deviation excess referred to gravity waves that took four hours, zero minutes and 1.2 seconds to travel between Maryland and Rochester. (p. 9)

Weber answered that he had never claimed that the 2.6-standard-deviation effect he had reported was a positive result. Nevertheless, by reporting a positive result where none was possible, Weber had cast further doubt on his own analysis procedures.

Garwin (1974; and Levine and Garwin, 1974) raised yet another question about Weber's results. They used a computer simulation to show that if Weber's apparatus was as he described it, then it could not have produced the result he claimed. They argued, in particular, that the narrow signal seen by Weber should have been broader (Figure 2.8).

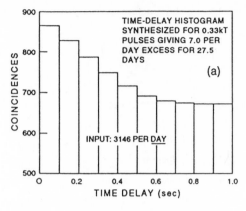

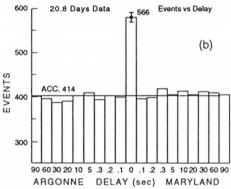

Figure 2.8. (a) Computer-simulation result obtained by Levine for signals passing through Weber's electronics. (b) Weber's reported result. The difference is clear. From Levine and Garwin (1974).

It seems clear that, according to the epistemological criteria discussed in the introduction, the critics' results were far more credible than Weber's. They had checked their results by independent confirmation, which included the sharing of data and analysis programs. They had also eliminated a plausible source of error, that of the pulses being longer than expected, by analyzing their results using the nonlinear algorithm and by looking for such long pulses. They had also calibrated their apparatuses by injecting known pulses of energy and observing the output.

In addition, Weber's reported result failed several tests suggested by these criteria. Weber had not eliminated the plausible error of a mistake in his computer program used to analyze the data. It was, in fact, shown that this error could account for his result. It was also argued that Weber's analysis procedure, which varied the threshold accepted, could also have produced his result. Having increased the credibility of his result when he showed that it disappeared when the signal from one of the two detectors was delayed, he then undermined it by obtaining a positive result when he thought two detectors were simultaneous, when, in fact, one of them had been delayed by four hours. As Garwin remarked, Weber's result itself also argued against its credibility. The coincidence in the time-delay graph was too narrow to have been produced by Weber's apparatus. Weber's analysis procedure also failed to detect calibration pulses.

The evidence against Weber's result was overwhelming. The discord was resolved by showing that the critics in their procedures had successfully applied the strategies outlined in the epistemology of experiment, whereas Weber's experiment had failed those epistemological criteria.

3

Millikan's Measurement of the Charge of the Electron

Millikan's oil-drop experiments are justly regarded as a major contribution to 20th-century physics (Millikan, 1911, 1913). They established the quantization of electric charge, the existence of a fundamental unit of charge, and also measured that unit of charge precisely. Earlier determinations of the charge of the electron had not established whether there was such a fundamental unit of electricity.[1] This was because previous experiments, which used a cloud of charged water droplets and observed the motion of the cloud under the influence of both gravity and an electric field and under gravity alone, measured the total charge of the cloud and could not therefore demonstrate that the value obtained was not a statistical average. Millikan was able to perform all of his measurements on solitary oil drops and thus avoid the difficulty. Examination of Millikan's laboratory notebooks reveals, however, that he was selective both in his choice of data and in his analysis procedure. But we shall see that the effects of this selectivity were small and did not significantly affect Millikan's final value for e, the charge of the electron. Nevertheless, Millikan's selectivity is problematic. Was he tuning both his data and his analysis procedure to get a desired result?

Millikan's Method

Let us briefly examine how Millikan demonstrated the existence of a fundamental unit of electrical charge and measured its value. Millikan allowed a single charged oil drop to fall a known distance in air. He did not measure the time of fall from rest, but allowed the drop to fall freely for a short distance before it passed a crosshair, which signaled the start of the time measurement. Because of air resistance, the drop was then traveling at a constant, terminal velocity. After the drop passed a second crosshair, which determined the time of fall at constant speed for the known distance between the crosshairs, an electric field was turned on. The charged oil drop then traveled upward at a different constant speed, and the time to ascend the same distance was measured. These two time measurements allowed the determination of both the mass of the drop and its total charge.

The equation of motion of an oil drop falling under an upward-directed electric field F is

$$m\ddot{x} = mg - K\dot{x} - e_n F,$$

where e_n is the drop's charge and m its mass compensated for the buoyant force of air. According to Stokes's law, which holds for a continuous retarding medium, $K = 6\pi a\mu$, where a is the drop's radius and μ the air's viscosity. To take into account the particulate character of air, Millikan replaced K by $K/(1 + b/pa)$, where p is the air pressure and b a parameter determined from the experimental data. Because all measurements were made at terminal velocities, the acceleration $\ddot{x}$ vanished (recalling that v_g is opposite in direction to v_f):

$$e_n = mg/F \left[(v_f + v_g)/v_g\right];$$

$$e_n = ne = \left[(9\pi d \sqrt{(2/g)})/F\right] \left[\mu^3/(\sigma - \rho)(1 + b/pa)^3\right]^{1/2}$$
$$\left[\sqrt{v_g} \left(1/t_g + 1/t_f\right)\right],$$

where e_n, the total charge on the drop, is assumed to be an integral multiple of a unit; d is the distance traveled, either up or down; and g is the acceleration of gravity.[2]

Not only did the charge on the oil drop sometimes change spontaneously due to absorption of charge from the air or ionization, but Millikan induced such changes with either a radioactive source or an x-

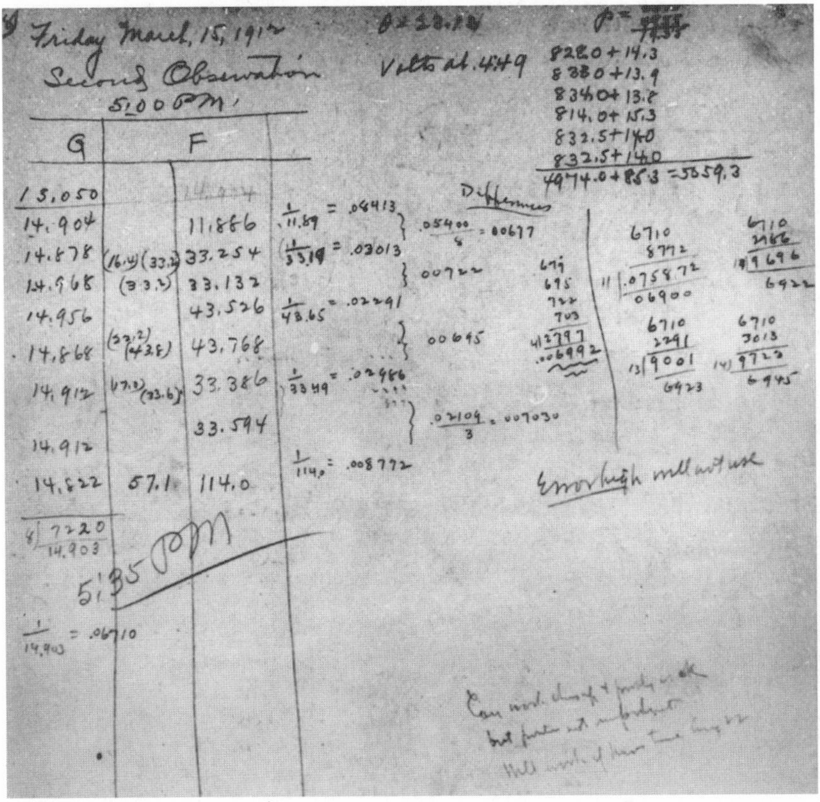

Figure 3.1. Millikan's data sheet for 15 March 1912 (second observation). Notice "Error high will not use" at the center-right. Courtesy California Institute of Technology Archives.

ray source. One can calculate the change in charge using successive times of ascent (before and after the change):

$$\Delta e_n = (\Delta n)e = [(9\pi d \sqrt{(2/g)})/F]\,[\mu^3/(\sigma - \rho)(1 + b/pa)^3]^{1/2}$$
$$[\sqrt{v_g}\,(1/t'_f - 1/t_f)],$$

where t'_f and t_f are the successive times of ascent. If both the total charge on the drop and the change in charge are multiples of some fundamental unit of charge, then $1/n(1/t_g + 1/t_f)$ should equal $(1/\Delta n)(1/t'_f - 1/t_f)$. Millikan could easily estimate $(1/t'_f - 1/t_f)$, because Δn was usually a very small number—often equal to one—for the smallest change in the charge of the drop.

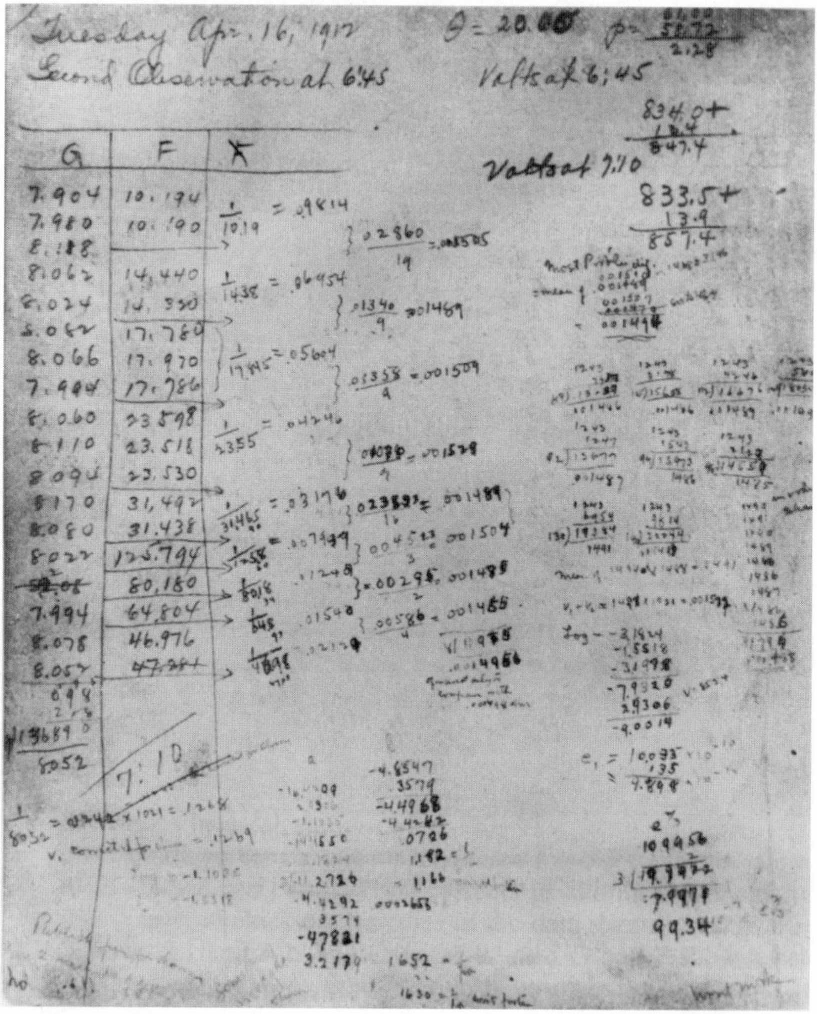

Figure 3.2. Millikan's data sheet for 16 April 1912 (second observation). Notice "Won't work" in lower right-hand corner. Courtesy California Institute of Technology Archives.

Since n' [our Δn] is always a small number and in some of the changes always had the value 1 or 2 its determination for any change is obviously never a matter of the slightest uncertainty. On the other hand n is often a large number, but with the aid of the known values of n' it can always be found with absolute certainty as long as it does not exceed say 100 or 150.[3] (Millikan, 1913, pp. 123–24)

Sample data sheets from Millikan's experiments are shown in Figures 3.1 and 3.2. (These sheets are from Millikan's notebooks of 1911 and 1912. The results of that experiment were published in 1913.) The columns labeled "G" and "F" are the measurements of t_g and t_f, respectively. The average value of t_g and its reciprocal are given at the bottom of column G. To the right of column F are calculations of $1/t_f$ and of $[(1/\Delta n)(1/t'_f - 1/t_f)]$. Further to the right is the calculation of $[(1/n)(1/t_f + 1/t_g)]$. The top of the page gives the date, the number and time of the observation, the temperature θ, the pressure p, and the voltage readings (which include the actual reading plus a correction), and the time at which the voltage was read. The data combined with the physical dimensions of the apparatus, the density of clock oil and of air, the viscosity of air, and the value of g are all that is required to calculate e.

Millikan's Results

Millikan could determine e from both the total charge of the drop and from the changes in the charge. Not only did these values agree very well, but the average value obtained from different drops was also the same. Millikan remarked, "The total number of changes which we have observed would be between one and two thousand, and *in not one single instance has there been any change which did not represent the advent upon the drop of one definite invariable quantity of electricity or a very small multiple of that quantity*" (Millikan, 1911, p. 360). For Millikan, and for most of the physics community, these results established the quantization of charge. The value that Millikan found in 1911 for the fundamental unit of charge, the charge on the electron, was 4.891×10^{-10} esu. (Millikan did not give a numerical uncertainty, but estimated the uncertainty as approximately 0.2 percent.)[4]

Following the completion of his 1911 paper, Millikan continued his oil-drop measurements. His intent was to improve both the accuracy and the precision of the measurement of e. He made improvements in his optical system and determined a better value for the viscosity of air. In addition, he took far more data in this second experiment. Millikan's new measurement gave a value of $e = (4.774 \pm 0.009) \times 10^{-10}$ esu.[5] This value differs considerably from his 1911 value of 4.891×10^{-10} esu:

The difference between these numbers and those originally found by the oil-drop method, $e = 4.891$, was due to the fact that this much more elaborate and prolonged study had the effect of changing every one of the three factors η [the viscosity of air], A [related to the correction parameter b in Stokes's law], and d [the distance between the crosshairs], in such a way as to lower e and to raise N [Avogadro's number]. The chief change, however, has been the elimination of faults of the original optical system. (Millikan 1913, pp. 140–41)

In producing his new value of e, Millikan engaged in selectivity in both the data he used and in his analysis procedure. In presenting his results in 1913, Millikan stated that the 58 drops under discussion had provided his entire set of data. *"It is to be remarked, too, that this is not a selected group of drops but represents all of the drops experimented upon during 60 consecutive days,* during which time the apparatus was taken down several times and set up anew" (Millikan 1913, p. 138). This is not correct. An examination of Millikan's notebooks for this period shows that Millikan took data for this measurement from 28 October 1911 to 16 April 1912.[6] My own count of the number of drops experimented on during this period is 175. Even if one were willing to count only those observations made after 13 February 1912, the date of the first observation Millikan published, there are 49 excluded drops: Of 107 drops experimented on between 13 February and 16 April, Millikan published only 58. We might suspect that Millikan selectively analyzed his data to support his preconceptions about both charge quantization and the value of e.[7]

Millikan's selectivity included suppressing all of the data for some drops, suppressing some of the data within the data set for a single drop, and choosing various methods of calculation. In discussing this selectivity, we should, however, remember that Millikan had far more data than he needed to improve the uncertainty in the measured value of e by approximately a factor of ten. He used only published drops—23 out of a total of 58—which had a correction due to Stokes's law of less than six percent to calculate his final value of e. This was to guard against any effect of an error in that correction.[8]

In experiments conducted before 13 February 1912, Millikan had labored to make his apparatus work properly. He was particularly worried about convection currents inside the device that could change the path of the oil drop. He made several tests on slow drops, for which convection effects would be most apparent. Millikan's comments on these tests are quite illuminating. On 19 December 1911, he remarked, "This work on a very slow drop was done to see whether there were appreciable convection currents. The results indicate that there were. Must look more carefully henceforth to tem[perature] of room."[9] On 20 December: "Conditions today were particularly good and results should be more than usually reliable. We kept tem very constant with fan, a precaution not heretofore taken in room 12 but found yesterday to be *quite* essential." On 9 February 1912, he disregarded his first drop because of uncertainty caused by convection; after the third drop he wrote, "This is good for so little a one but on these very small ones I must avoid convection still better." No further convection tests are recorded. By 13 February, it seems that the device was working to Millikan's satisfaction, because he eventually published data from the very first drop recorded on that day. The data from 68 drops taken before 13 February were omitted from publication, because Millikan was not convinced that his experimental apparatus had been working properly. It was not producing "good" data.[10]

After this date, we must assume that the apparatus was working properly unless we are explicitly told otherwise. There are 107 drops in question, of which 58 were published. Millikan made no calculation of *e* on 22 of the 49 unpublished drops. The most plausible explanation for why Millikan did not do so is that when he performed his final calculations in August 1912, the drops seemed superfluous: He did not need more drops for the determination of *e*.

The 27 events that Millikan did not publish and for which he calculated a value of *e* are more worrisome. Millikan knew the results he was excluding. Twelve of these were excluded from the set of published drops because they seemed to require a second-order correction to Stokes's law. These were very small drops, for which the value of Millikan's correction to Stokes's law, b/pa, was larger than one. This made Millikan's use of a perturbation series expansion of Stokes's law for those drops very questionable. There is no easy way to calculate the correction for such drops, so Millikan, having so much data, decided to exclude them.[11] Of the remaining 15 calculated events, Millikan excluded two because the apparatus was not working properly, five because there was insufficient data to make a reliable determination of *e*, and two for no apparent reason. We are left with six drops. One is quite anomalous.[12] In the five remaining

Table 3.1.

Comparison of Millikan's and Franklin's values of e

| | e (×10^{10} esu) | | σ[a] | |
Drops Used in Calculation	Millikan	Franklin	Millikan	Franklin
First 23[b]	4.778[c]	4.773	±0.002	±0.004
All 58	4.780	4.777	±0.002	±0.003
Almost all drops[d]	4.781	4.780	±0.003	±0.003

[a]Statistical error in the mean.
[b]These are the events that Millikan used to determine e.
[c]Although Millikan used a value $\mu = 0.001825$ for the viscosity of air in almost all of his calculations, in reporting his final value for e he used $\mu = 0.001824$. This accounts for the change from 4.774 to 4.778 in Millikan's final value. To make the most accurate comparison, I used $\mu = 0.001825$ in all my recalculations.
[d]This includes the 58 published drops, 25 unpublished drops measured after 13 February 1912, and some small corrections. For details, see Franklin (1981).

cases, Millikan not only calculated a value for e but compared it with an expected value. The four earliest events have values of pa that would place them in the group Millikan used to determine e. His only evident reason for rejecting these five events is that their values did not agree with his expectations. Including these events among the 23 that Millikan used to determine e would not significantly change the average value of e, but would increase the statistical error of the measurement very slightly (see Table 3.1).

In addition to excluding these five drops from publication, Millikan's cosmetic surgery touched 30 of the 58 published events. As shown in Figures 3.1 and 3.2, Millikan made many measurements of the time of fall under gravity and of the time of ascent with the electric field on for each drop. In the data set for each of these 30 drops, Millikan excluded one or more (usually less than three) of these measurements. This group of 30 drops included several of the 23 drops used by Millikan in his final determination of e as well as some of the 35 published drops that were not used. My recalculation of these events, using all of the data for each drop, gives results little different from Millikan's. His exclusion of these measurements was not based on the value of e he obtained for the drops, because, in general, Millikan did not include these measurements in his calculations, and therefore did not know their effect.

As discussed earlier, there are two ways to calculate e from the oil-drop data. The first uses the total charge on the drop, whereas the second uses the changes in charge. Millikan claimed that he had used the first method exclusively because the large number of measurements of t_g provided a more accurate determination of e. In at least 19 of the 58 published events,[13] however, he used either the average value of the two methods, some combination of the two that is not a strict average, or the second method alone. (Figure 3.1 shows the data sheet for an event that Millikan excluded because he thought the difference in the value of e obtained by the two different methods was too large.) In general, the effects are small and the result of his tinkering, once again, is to reduce the statistical error very slightly rather than to change the mean value of e.

The effect of all of Millikan's selectivity is shown in Table 3.1. This includes Millikan's results along with my own recalculation of his data. The results of his selectivity are quite small.

Discussion

What can we conclude from this episode? Millikan intended to establish the quantization of charge and to measure the fundamental unit more accurately and precisely than had been done previously. It is apparent that he succeeded: There is no reason to disagree with his assessment that, in 1913, there was "no determination of e . . . by any other method which does not involve an uncertainty at least 16 times as great as that represented in these measurements." His apparently arbitrary exclusion of five drops for which he had calculated e, a possible worry, had, as we have seen, an utterly negligible effect on his final result. Because Millikan knew the value of e obtained from the events he was excluding, he also knew that the effect of the exclusions and of his selective calculations on his final result was small.[14]

Nevertheless there is strong evidence that Millikan tuned his cuts and his analysis procedure to obtain the result he wanted. Several of these cuts seem quite legitimate: the exclusion of the early drops because he was not sure that the experimental apparatus was working properly; the exclusion of some data within the set for a drop; and the exclusion of later events because he simply did not need them for his calculations. The exclusion of drops for which he calculated a value of e and could thus select the value he wanted as well as his choice of calculational method are not justified. The physics community did not know that Millikan tuned his results. Unlike the episodes of the measurement of the K_{e2}^+ branching ratio (Chapter 1) and of Weber's attempts to detect gravity waves (Chapter 2), in which the cuts were publicly accessible,[15] Millikan's questionable selectivity remained private. In fact, his statements in his 1913 paper that the drops were not selected and that he used only one method of calculation seem to have been designed to conceal that selectivity.

What then are the safeguards against procedures such as Millikan's, which, in less sure hands, could easily have unfortunate results? In this case the answer is replication. The value of e is an important physical quantity. It is used in the calculation or determination of many important physical constants (e.g., Avogadro's number, the Rydberg constant). There were many repetitions of Millikan's measurement.[16] Had Millikan's selectivity grossly affected his measured value of e, there would certainly have been a discrepancy with later measurements.

4

The Disappearing Particle
The Case of the 17-keV Neutrino

In Chapter 2, I discussed the history of the early attempts to detect gravitational radiation. We saw how different choices of analysis procedures by Weber and his critics led to discordant experimental results. The issues of selectivity and discord are also linked in the recent history concerning the possible existence of a heavy, 17-keV neutrino.[1] In this case, the problems were more complex, because—unlike the case of gravity waves, in which only Weber and his collaborators obtained positive results—results on both sides of the issue were reported by several groups. In addition, both the original positive claim and all subsequent positive claims were obtained in experiments using one type of apparatus, namely, that incorporating a solid-state detector, whereas the initial negative evidence resulted from experiments using another type of detector (a magnetic spectrometer).[2] These were both seemingly reliable types of experimental apparatus. Solid-state detectors had been in wide use since the early 1960s, and their functional principles were understood. Magnetic spectrometers had been used in nuclear β-decay experiments since the 1930s and both the problems and advantages of using this technique had been well studied.[3] This episode is an illustration of discordant results ob-

For details of this episode, see Franklin (1995a).

tained using different types of apparatus. One might worry that the discordant results were due to some crucial difference between the types of apparatus or to different sources of background that might mimic or mask the signal. There were also questions of selectivity concerning the proper theoretical model with which to compare the experimental data and the appropriate energy range for that comparison.

The Appearance

The 17-keV (or heavy) neutrino was "discovered" by Simpson (1985). He had searched for a heavy neutrino by looking for a kink in the decay-energy spectrum or in the Kurie plot[4] at an energy equal to the maximum-allowed decay energy minus the mass of the heavy neutrino, in energy units. The fractional deviation in the Kurie plot value $\Delta K/K$ is approximately $R[1 - M_2^2/(Q - E)^2]^{1/2}$, where M_2 is the mass of the heavy neutrino, R is the intensity of the second neutrino branch, Q is the total energy available for the transition, and E is the energy of the electron. Simpson's initial experimental result for the decay-energy spectrum of tritium is shown in Figure 4.1. A kink—a marked change in the slope of the $\Delta K/K$ graph—is clearly seen at electron energy $T_\beta = 1.5$ keV, corresponding to a 17-keV neutrino. (The maximum decay energy for tritium is 18.6 keV. If there were no effect due to the presence of a heavy neutrino, this graph would be a horizontal straight line.) "In summary, the β spectrum of tritium recorded in the present experiment is consistent with the emission of a heavy neutrino of mass about 17.1 keV and a mixing probability [the fraction of heavy neutrinos] of about 3%" (Simpson, 1985, p. 1893).[5]

Within a year there were five attempted replications of Simpson's experiment (Altzitzoglou et al., 1985; Apalikov et al., 1985; Datar et al., 1985; Markey and Boehm, 1985; Ohi et al., 1985). Each of them gave negative results. The experiments set limits of <1% for a 17-keV branch of the decay, in contrast with Simpson's value of 3%. A typical result, that of Ohi et al. (1985), is shown in Figure 4.2 and should be compared with Simpson's result shown in Figure 4.1. No kink of any kind is apparent in the experimental data.

Each of the subsequent experiments had examined the beta-decay spectrum of ^{35}S, and searched for a kink at an energy of 150 keV, 17 keV below the endpoint energy of 167 keV. Three of the experiments—those

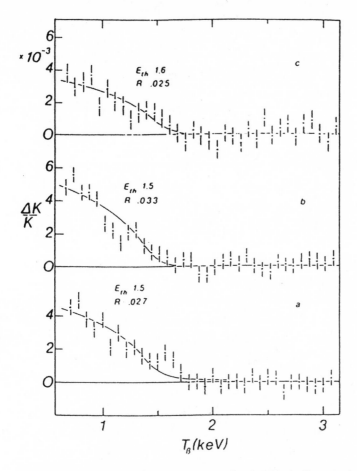

Figure 4.1. Data from three runs presented as $\Delta K/K$ (the fractional change in the Kurie plot) as a function of the kinetic energy of the β particles. E_{th} is the threshold energy (the difference between the endpoint energy and the mass of the heavy neutrino). A kink is clearly seen at $E_{th} = 1.5$ keV, or at a mass of 17.1 keV. Run *a* included active pile-up rejection, whereas runs *b* and *c* did not. Run *c* was the same as *b* except that the detector was housed in a soundproof box. No difference is apparent. From Simpson (1985).

of Altzitzoglou et al. (1985), Apalikov et al. (1985), and Markey and Boehm (1985)—used magnetic spectrometers. Those of Datar et al. (1985) and Ohi et al. (1985) used Si(Li) detectors, the same type used by Simpson. In the latter two cases, however, the source was not implanted in the detector, as Simpson had done, but was separated from it. Such an arrangement would change the atomic-physics correc-

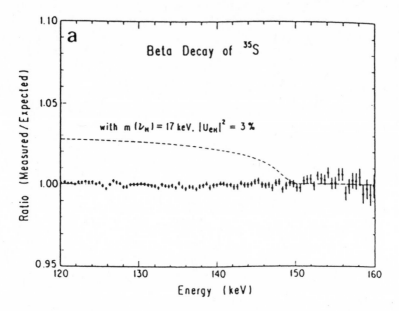

Figure 4.2. The ratio of the measured ^{35}S beta-ray spectrum to the theoretical spectrum. A 3% mixing of a 17-keV neutrino should distort the spectrum as indicated by the dashed curve. From Ohi et al. (1985).

tions to the spectrum.[6] In addition, the ^{35}S beta-decay sources used in the experiments had a higher endpoint energy than did the tritium used by Simpson (167 keV in contrast to 18.6 keV). This higher endpoint energy made the atomic-physics corrections to the beta-decay spectrum less important.

Simpson's first report of the 17-keV neutrino was unexpected. It was not predicted, or even suggested, by any existing theory. Faced with such a startling result, the physics community took a reasonable approach. Some theoretical physicists tried to explain the result within the context of accepted theory. They argued that a plausible alternative explanation of the result had not been considered, raising the question of whether the correct theory had been used to analyze the data and compare the experimental result with the theory of the phenomenon. This is an important point: An experimental result is not usually immediately given by an examination of the raw data, but requires considerable analysis. In this case, the analysis included atomic-physics corrections, needed for the comparison of the theoretical spectrum and the experimental data.

Everyone involved agreed that such corrections had to be made. The question was: What were the proper corrections? The atomic-physics corrections used by Simpson in his analysis, particularly the screening potential, were questioned by others (Haxton, 1985; Kalbfleisch and Milton, 1985; Drukarev and Strikman, 1986; Eman and Tadic, 1986; Lindhard and Hansen, 1986).[7] These suggestions were aimed at accommodating the unexpected result within the accepted theory. Several calculations indicated, at least qualitatively, that Simpson's result could be accommodated, and that there was no need for a new particle. "A detailed account of the decay energy and Coulomb-screening effects raises the theoretical curve in precisely this energy range so that little, if any, of the excess remains" (Lindhard and Hansen, 1986, p. 965). Thus there was a question of selectivity in the choice of atomic-physics corrections.

The combination of negative experimental searches combined with plausible theoretical explanations of Simpson's result had a chilling effect on the topic. Almost all experimental work on the subject ceased. Simpson, however, continued his work. He presented further evidence in support of the 17-keV neutrino, using a somewhat modified experimental apparatus (Simpson, 1986b). He also took the criticism of his work seriously and presented an analysis of his new data that incorporated the screening potential suggested by his critics. Although this reduced the size of his effect by approximately 20%, the effect was still clearly present. Simpson had shown that his result was reasonably robust under variations in the atomic-physics correction to the decay spectrum. He also questioned whether the analysis procedures used in the five negative searches were adequate to set the upper limits they had reported. He argued that the wide energy range used to fit the β-decay spectrum tended to minimize any possible effect of a heavy neutrino, which would appear primarily in a narrow energy band near the threshold. He also questioned the procedure of merely adding the contribution of a 17-keV neutrino to the already-fitted spectrum, a point with which others agreed (Borge et al., 1986). Simpson further questioned whether the "shape-correction" factor needed to fit the spectra in magnetic spectrometer experiments could mask a kink due to the presence of a heavy neutrino. These questions would have to be answered. Simpson (1986a) also presented a reanalysis of Ohi's data using his own preferred analysis procedure. He found a positive effect (Figure 4.3). (Compare this figure with Ohi's own reported result, shown in Figure 4.2. How this conflict arose is discussed later in this chapter.)

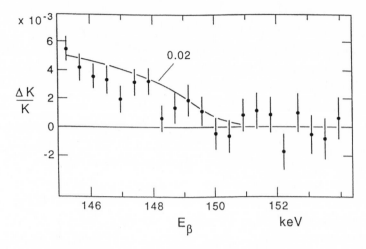

Figure 4.3. $\Delta K/K$ for the ^{35}S spectra of Ohi et al. (1985) as recalculated by Simpson. From Simpson (1986a).

Further negative evidence was provided by Borge et al. (1986), Hetherington et al. (1986, 1987), and Zlimen et al. (1988). Hetherington et al. urged caution concerning Simpson's method of data analysis. They pointed out that "concentrating on too narrow a region can lead to misinterpretation of a local statistical anomaly as a more general trend" (Hetherington et al., 1987, p. 1512). The issue of the appropriate energy range for the analysis had not yet been decided. At the end of 1988, the situation was much as it had been at the end of 1985. Simpson had presented positive results on a 17-keV neutrino. There were nine negative experimental reports as well as plausible theoretical explanations of his result.

In 1989, Simpson and Hime presented two additional positive results, using both tritium and ^{35}S, the spectrum used in the original negative searches. In these reports, the value of the mixing fraction of 17-keV neutrinos had been reduced to approximately 1%. The new atomic-physics corrections had reduced the originally reported effect from 3% to 1.6% (Hime and Simpson, 1989; Simpson and Hime, 1989).

The situation changed dramatically in 1991. New positive results were reported at both conferences and in the published literature by groups at Oxford (Hime and Jelley, 1991) and at Berkeley (Sur et al., 1991).[8] These results were quite persuasive. Hime and Jelley had incorporated antiscatter baffles into their apparatus to guard against a distortion of the spectrum caused by scattering of the decay electrons, a possible problem in the earlier experiments. The Berkeley group had embedded their ^{14}C source in a solid-state detector and included a guard ring veto to reject decays occurring near the boundary, which might not deposit their full energy and thus distort the spectrum. They claimed that their result "supports the claim by Simpson that there is a 17-keV neutrino emitted with ~1% probability in β decay" (Sur et al., 1991, p. 2447). They also claimed to rule out the null hypothesis (no heavy neutrino) at the 99% confidence level. A further positive result was reported by Zlimen et al. (1991). These new results generated considerable new experimental and theoretical work. Sheldon Glashow, a Nobel-Prize-winning theorist, remarked, "Simpson's extraordinary finding proves that Nature's bag of tricks is not empty, and demonstrates the virtue of consulting her, not her prophets" (Glashow, 1991, p. 257).

The Disappearance

The summer of 1991 marked the high point in the life of the 17-keV neutrino. From that time forward, only negative results would be reported, and errors would be found in the most persuasive positive results. Piilonen and Abashian (1992) suggested that Hime and Jelley had overlooked a background effect that might have simulated the effect of a 17-keV neutrino in their experiment. The appearance of several negative results (discussed later in this chapter) encouraged Hime (1993) to consider the Piilonen-Abashian suggestion seriously and to reanalyze his own result. He found, using an experimentally checked Monte Carlo calculation, that the scattering of the decay electrons in the experimental apparatus could explain the result without the need for a 17-keV neutrino. "It will be shown that scattering effects are sufficient to describe the Oxford β-decay measurements and that the model can be verified using existing calibration data. Surprisingly, the β spectra are very sensitive to the small corrections considered" (Hime, 1993, p. 166). He also suggested that similar effects might explain his earlier positive results obtained in collaboration with Simpson.

Hime briefly reviewed the evidential situation, noting that the major evidence against the existence of the 17-keV neutrino came from magnetic spectrometer experiments, in which questions had been raised concerning the shape corrections. He commented that Bonvicini (in a CERN report, CERN-PPE/92-54, published later as Bonvicini [1993]) had shown that nonlinear distortions could mask the presence of a heavy-neutrino signature and still be described by a smooth shape correction.[9] He remarked, however, that "A measurement of the ^{63}Ni spectrum (Kawakami, Kato et al., 1992) has circumvented this difficulty. The sufficiently narrow energy interval studied, and the very high statistics accumulated in the region of interest, makes it very unlikely that a 17-keV threshold has been missed in this experiment" (Bonvicini, 1993, p. 165). He also cited a new result from a group at Argonne National Laboratory (Mortara et al. [1993], discussed in detail in the next paragraph), that provided "convincing evidence against a 17-keV neutrino." In particular, the Argonne group had demonstrated the sensitivity of their magnetic spectrometer experiment to a possible 17-keV neutrino by admixing a small component of ^{14}C in their ^{35}S source and detecting the resulting kink in their composite spectrum. These negative results provided the impetus for Hime's reexamination of his result.

Some of the evidence that Hime cited against the 17-keV neutrino was provided by the Argonne group (Mortara et al., 1993). This experiment used a solid-state Si(Li) detector (the same type originally used by Simpson), an external ^{35}S source, and a solenoidal magnetic field to focus the decay electrons. The field also had the effect of reducing the backscattering of the decay electrons, a possible problem when interpreting the data. Their final result for the mixing probability of the 17-keV neutrino, shown in Figure 4.4, was $\sin^2\theta = -0.0004 \pm 0.0008$ (statistical) ± 0.0008 (systematic). They had found no evidence for a 17-keV neutrino.

The experimenters demonstrated the sensitivity of their apparatus to a possible 17-keV neutrino:

To assess the reliability of our procedure, we introduced a known distortion into the ^{35}S beta spectrum and attempted to detect it. A drop of ^{14}C-doped valine ($E_0 - m_e \sim 156$ keV) was deposited on a carbon foil and a much stronger ^{35}S source was deposited over it. The data from the composite source were fitted using the ^{35}S theory, ignoring the ^{14}C contaminant. The residuals are shown in Figure [4.5]. The distribution is not flat; the solid curve shows the expected deviations from the single component spectrum with the measured amount of ^{14}C. The fraction of decays from ^{14}C determined from the fit to the beta spectrum is $(1.4 \pm 0.1)\%$. This agrees with the value of 1.34% inferred from measuring the total

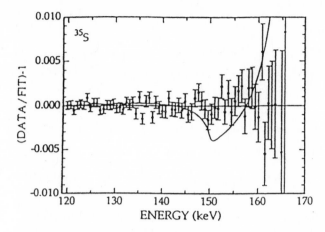

Figure 4.4. Residuals from a fit to the pile-up corrected ^{35}S data, assuming no massive neutrino; the reduced χ^2 for the fit is 0.88. The solid curve represents the residuals expected for decay with a 17-keV neutrino and $\sin^2\theta = 0.85\%$; the reduced χ^2 of the data is 2.82. From Mortara et al. (1993).

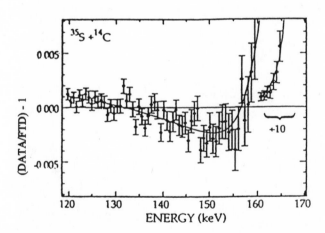

Figure 4.5. Residuals from fitting the beta spectrum of a mixed source of ^{14}C and ^{35}S with a pure ^{35}S shape; the reduced χ^2 of the data is 3.59. The solid curve indicates residuals expected from the known ^{14}C contamination. The best fit yields a mixing of $(1.4 \pm 0.1)\%$ and reduced χ^2 of 1.06. From Mortara et al. (1993).

decay rate of the ^{14}C alone while the source was being prepared. This exercise demonstrates that our method is sensitive to a distortion at the level of the positive experiments. Indeed, the smoother distortion with the composite source is more difficult to detect than the discontinuity expected from the massive neutrino.

In conclusion, we have performed a solid-state counter search for a 17 keV neutrino with an apparatus with demonstrated sensitivity. We find no evidence for a heavy neutrino, in serious conflict with some previous experiments. (Mortara et al., 1993, p. 396)

At this time, the Berkeley group began to question their own positive result. Further experimental runs had shown that the guard ring seemed to generate a spurious 17-keV neutrino signal. They searched for the cause of the artifact in their ^{14}C data. The cause, found in 1993, was quite subtle. The center detector was separated from the guard ring by cutting a groove in the detector:

The n^{+} is divided by a 1-mm-wide circular groove into a "center region" 3.2 cm in diameter, and an outer "guard ring." By operating the guard ring in anticoincidence mode, one can reject events occurring near the boundary which are not fully contained within the center region. (Sur et al., 1991, p. 2444).

Such events would not give a full energy signal and would thus distort the observed spectrum.

What the Berkeley group found was that ^{14}C decays occurring under the groove shared the energy between both regions without necessarily giving a veto signal, and thus gave an incorrect event energy, distorting the spectrum. They also found that, although their earlier tests had indicated that the ^{14}C was uniformly distributed in the detector, their new tests showed that between one-third and one-half of the ^{14}C was localized in grains. They also found that approximately 1% of the grains were located under the groove. Thus the localization of the ^{14}C combined with the energy sharing gave rise to a distortion of the spectrum, simulating that expected from a 17-keV neutrino (Norman, pers. comm., LBL-36136, 1994; Wietfeldt et al., 1993).

The newer negative results were persuasive not only because of their improved statistical accuracy, but also because they were able to demonstrate that their experimental apparatuses could detect a kink in the spectrum if one were present. This was a direct experimental check that there were no effects present that would mask the presence of a heavy neutrino.

These experiments met Hime's suggested criteria of a demonstrated ability to detect a kink combined with high statistics so that a local analysis of the spectrum could be done.[10] Ohshima et al. had also shown that the shape-correction factors used in their experiment did not mask any possible 17-keV neutrino effect (Kawakami et al., 1992; Ohshima, 1993; Ohshima et al., 1993).[11] This combination of persuasive evidence against the existence of a 17-keV neutrino and the demonstrated and admitted problems with the positive results decided the issue. There was no 17-keV neutrino.

It seems clear that this decision was based on experimental evidence, discussion, and criticism—in other words, on epistemological criteria. It had been shown that the two most persuasive positive results had overlooked effects that mimicked the presence of a 17-keV neutrino. The Sherlock Holmes strategy had been incorrectly applied. In addition, the new negative results had answered the criticisms made previously concerning the "shape-correction" factor and had demonstrated that the experiments could detect a kink in the spectrum if one were present. What, then, of the question of selectivity? The careful reader may still be asking why Simpson's reanalysis of Ohi's data gave such different results. This is discussed in the next section.

Selectivity and the 17-keV Neutrino

One aspect of the selectivity issue in this episode is the choice of the energy range used to fit the decay-energy spectrum, so that the experimental and theoretical spectra could be compared. Simpson (1986b) had argued that because 45% of the expected effect occurred within 2 keV of the neutrino threshold, a narrow energy range around that threshold should be used:

in trying to fit a very large portion of the β spectrum, the danger that slowly-varying distortions of a few percent could bury a threshold effect seems to have been disregarded. One cannot emphasize too strongly how delicate is the analysis when searching for a small branch of a heavy neutrino, and how sensitive the result may be to apparently innocuous assumptions. (p. 576)

Hetherington et al. (1987) suggested caution:

It has been argued [by Simpson] that in order to avoid systematic errors, only a narrow portion of the beta spectrum should be employed in looking for the threshold effect produced by heavy neutrino mixing. If one accepts this argument, our data in the narrow scan region set an upper limit of 0.44% [much lower than the 3% effect originally found by Simpson]. However, we feel that concentrating on a narrow region and excluding the rest of the data is not warranted provided adequate care is taken to account for systematic errors. The rest of the spectrum plays an essential role in pinning down other parameters such as the endpoint. Furthermore, concentrating on too narrow a region can lead to misinterpretation of a local statistical anomaly as a more general trend which, if extrapolated outside the region, would diverge rapidly from the actual data. (p. 1512)[12]

This issue was dramatically demonstrated by Simpson's reanalysis (1986a) of the data of Ohi et al. (1985). Recall from Figure 4.2 that their result showed no evidence for a 17-keV neutrino. Simpson's reanalysis of that same data shows clear positive evidence (Figure 4.3). How can the same data provide both positive and negative evidence for the same effect? The answer is that the analysis procedures were quite different. Ohi and collaborators had used a wide energy range for their analysis.[13] As Morrison (1992) later showed, the positive effect found by Simpson was due to his use of a narrow energy range for his reanalysis of Ohi's data:

The question then is, How could the apparently negative evidence of Figure [4.2] become the positive evidence of Figure [4.3]? The explanation is given in Figure [4.6], where a part of the spectrum near 150 keV is enlarged. Dr. Simpson only considered the region 150 keV ± 4 keV (or more exactly +4.1 and –4.9 keV). The procedure was to fit a straight line, shown solid, through the points in the 4 keV interval above 150 keV, and then to make this the base-line by rotating it down through about 20° to make it horizontal. This had the effect of making the points in the interval 4 keV below 150 keV appear above the extrapolated dotted line. This, however, creates some problems, as it appears that a small statistical fluctuation between 151 and 154 keV is being used: the neighboring points between 154 and 167, and below 145 keV, are being neglected although they are many standard deviations away from the fitted line. [Simpson's straight-line fit to the data just above 150 keV and its extrapolation is the line going from lower left to upper right. Comparing the data points to this line generates the positive effect seen in Figure 4.6. The dotted curve above the data is the effect expected for a heavy neutrino.] Furthermore, it is important, when analyzing any data, to make sure that the fitted curve passes through the end-point of about 167 keV, which it clearly does not. (p. 600)

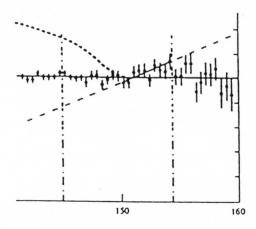

Figure 4.6. Morrison's reanalysis of Simpson's reanalysis of Ohi's result. From Morrison (1992).

150   160

The caution urged by both Hetherington and collaborators and by Hime was justified.[14]

How was this issue dealt with? Several later experiments used *both* a narrow and a wide energy range in the analysis of their data (Hetherington et al., 1987; Kawakami et al., 1992; Radcliffe et al., 1992; Ohshima, 1993; Ohshima et al., 1993). For example, Hetherington et al. (1987) concluded that their results from both the wide- and narrow-energy range analyses agreed and that "The shape of the plot and the reduced χ^2 value clearly rule out this large a mixing fraction [3%] for the 17 keV neutrino" (p. 1510).

Ultimately, the decision that the 17-keV neutrino did not exist was based on finding errors in the two most persuasive positive results and by the overwhelming negative evidence provided by experiments that deliberately avoided the data analysis issues posed by the narrow vs. the wide energy range. The first of these experiments was that of a Tokyo group (Kawakami et al., 1992; Ohshima, 1993; Ohshima et al., 1993). These experimenters noted some of the problems that plague experiments using wide energy regions and decided therefore to concentrate "on performing a measurement of high statistical accuracy, in a narrow energy region, using very fine energy steps. Such a restricted energy scan . . . also reduced the degree of energy-dependent corrections and other related systematic uncertainties" (Kawakami et al., 1992, p. 45). The data were taken over three overlapping energy ranges: 41.2–46.3 keV, 45.7–51.1 keV, and 50.5–56.2 keV (the threshold for a 17-keV neutrino occurs at

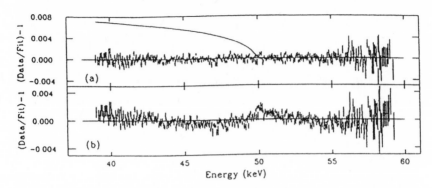

Figure 4.7. Deviations from the best global fit with: (a) $|U|^2$ free; (b) $|U|^2$ fixed to 1%. The curve in (a) indicates the size of a 1% mixing effect of the 17-keV neutrino. From Ohshima (1993).

approximately 50 keV). The results are shown in Figure 4.7, for (a) the mixing probability as a free parameter, and (b) with the probability fixed at 1%. The effect expected for a 17-keV neutrino with a 1% mixing probability is also shown in (a). No effect is seen. Their best value for the mixing probability of a 17-keV neutrino was [−0.011 ± 0.033 (statistical) ± 0.030 (systematic)]%, with an upper limit for the mixing probability of 0.073% at the 95% confidence level. This was the most stringent limit yet and was certainly far lower than the approximately 1% effect found by Simpson and others. "The result clearly excludes neutrinos with $|U|^2 \geqslant$ 0.1% for the mass range 11 to 24 keV" (Ohshima, 1993, p. 1128).

Although the experiment's narrow energy range was designed to minimize the dependence of the result on the shape correction,[15] the experimenters also checked on the sensitivity of their result to that correction. They normalized their data in the three energy regions using the counts in the overlapping regions, and divided their data into two parts: (1) below 50 keV, which would be sensitive to the presence of a 17-keV neutrino, and (2) above 50 keV, which would not. They then fit their data in region (2) and extrapolated the fit to region (1). The resulting fit was far better than one that included a 1% mixture of the 17-keV neutrino, demonstrating that the shape correction was not masking a possible effect of a heavy neutrino. Bonvicini noted that this experiment, with its very high statistics, had answered essentially all of his criticism of spectrometer experiments convincingly. "Thus, I conclude that this experiment could not possibly have missed the kink and obtain[ed] a good χ^2 at the same time, in the case of an unlucky misfit of the shape factor" (Bonvicini, 1993, p. 115).[16] The Tokyo group had shown that their result was robust against changes in the energy range used in their analysis and was not sensitive to their use of a "shape-correction" factor.

The second experiment that provided decisive evidence against the existence of the 17-keV neutrino was that of a group at the Argonne National Laboratory (Mortara, Ahmad et al., 1993). Not only did this experiment find no evidence for a 17-keV neutrino (Figure 4.4),[17] but the experimenters had demonstrated that they would have observed such evidence had it been present.[18] That an effect of both the right size and shape was observed when the spectra of ^{35}S and ^{14}C were mixed demonstrated that their analysis procedure did not mask a real effect. The failure to observe any effect in the ^{35}S spectrum alone showed that the analysis procedure did not create an artifact that would mimic the effect of a heavy neutrino.

Discussion

The case of the 17-keV neutrino is yet another illustration of selectivity applied to analysis procedures. This selectivity was an important factor in producing the discordant results. The problem of the appropriate analysis procedure—in particular, the energy range to be used in the analysis—was resolved, in part, by several experiments in which both procedures were used and in which the results were robust under changes in the procedures. The two decisive, negative results (those of the Tokyo and Argonne groups) both demonstrated, albeit in different ways, that the choice of analysis procedure was not a relevant issue in assessing the validity of their results. The Tokyo group not only used both procedures, but had such high statistics that a kink in the spectrum of the size found by Simpson could not have been missed in their analysis. The Argonne group demonstrated that their experiment (apparatus and analysis procedure) would have detected a 17-keV neutrino at the 1% level had one been present. As we have seen earlier, the issue of discordant results was resolved by epistemologically convincing negative results combined with finding error in the positive results.[19]

5

Are There Really Low-Mass Electron-Positron States?

In the chapters on gravity waves and on the 17-keV neutrino, we saw that selectivity can be a cause of discordant experimental results. This was also the case in the recent controversy concerning the possible existence of low-mass electron-positron states produced in high-energy heavy ion-atom collisions. In this episode, we see selectivity applied to both data and to the procedures used to analyze that data. This episode is more complex than those discussed earlier because it includes results that could be replicated only some of the time and experiments performed under seemingly identical conditions that gave different results. The questions of what constitutes an adequate replication of an experiment and what are the "same" results was extremely important. In addition, even though there were problems with replicating them, the earliest results were all thought to be in sufficient agreement to support the existence of the electron-positron states, or at least to merit further investigation. The discord was recognized only later. Eventually the original results were shown to be incorrect: The consensus is that there are no low-mass electron-positron states.

Here I briefly outline the history. Early experiments found evidence that positrons with discrete energies, or positron lines, were produced in

For more details of this episode, see Franklin (1998).

high-energy heavy ion-atom collisions. Later experiments also detected an electron produced in coincidence with the positron and found peaks in the sum-energy spectrum ($E_{e+} + E_{e-}$), suggestive evidence for low-mass electron-positron states or particles.[1] Because the effects appeared for various pairs of projectiles and target nuclei, as well as in three different experimental apparatuses, the results had credibility, although there were some problems concerning their reproducibility. This credibility led others to further investigate the phenomena. The later experiments produced conflicting claims. Part of the difficulty in attempting to resolve this issue was that the results were not reproducible. Even when later experiments were done under very similar conditions to the original experiments, the effects did not always appear, and when they did, the results were not always identical. The failure to reproduce the effects cast doubt on the original results.

A recent analysis by Ganz and the EPOS II (Electron POsitron Solenoidal spectrometer) group (Ganz et al., 1996) has suggested that the observed effects might be artifacts created by the selection criteria used to produce the experimental result.[2] Applying certain selection cuts to one-half of his data and tuning the cuts to produce a maximal effect, Ganz and company found rather strong evidence (a five-standard-deviation effect) for a low-mass state. Applying identical cuts to the other half of his data showed no such evidence.[3] This failure cast doubt on the positive result and also on the analysis procedures initially used. The originally observed effect was five standard deviations above background. If the observation is a real effect, then it is very improbable that it would disappear in the analysis of the other half of the data. The implication is that the effect is an artifact created by the cuts. Although most physicists working in the field believe that the proposed low-mass electron-positron states do not exist, not everyone agrees (see Taubes, 1997; and note 12).

This episode nicely raises the question of whether particular experimental cuts or selection criteria can create an effect that is not really present. I examine how the physicists involved dealt with this issue and with the associated problem of discordant experimental results. The situation is made more difficult by the fact that the heavy ion-atom systems under investigation contain a large number of particles with many possible interactions. That one may study these reactions as a function of no fewer than fourteen different variables further increases the difficulty of the experiment. On the theoretical side, the situation was made difficult because (1) the heavy ion-atom systems being studied are quite complex; and (2) positrons may be created both by electromagnetic processes in the strong electric fields produced in heavy ion-atom collisions and by nuclear interactions.

Positron Line Spectra

The history of possible low-mass electron-positron states began with the demonstration that positrons were indeed produced in high-energy heavy ion-atom collisions (Backe et al., 1978; Kozhuharov et al., 1979). The motivation for these experiments was the theoretical speculation that positrons would be produced in the strong electric fields produced by such collisions. The positrons were expected to be produced in super-critical systems, those for which the binding energy of the lowest energy state is greater than $2m_e c^2$, where m_e is the mass of the electron. This is true for heavy ion-atom systems that have a total nuclear charge $Z_u > 173$. These early experiments, all performed at the heavy-ion accelerator at the Gesellschaft für Schwerionforschung (GSI), Darmstadt, Germany, showed a surprising enhancement of positron production at energies <400 keV. They were followed by more detailed experiments, which pro-duced unexpected results on the basis of accepted theory. The early experimental results illustrate quite clearly the problems of reproducibil-ity and sensitivity of experimental results to both the selection criteria and the experimental conditions. These would be continuing problems in the subsequent history.

The three early experiments—performed by the EPOS I, TORI, and the Orange groups—used two different methods for high-efficiency positron detection. Subsequently, no questions were raised concerning the adequacy of the different detectors. Everyone agreed that they were good positron detectors and that they had been carefully calibrated.

One of the intriguing new results found by the EPOS I group was the measurement for $^{238}U + ^{238}U$ at $E(^{238}U) = 5.9$ MeV/u (u is one atomic mass unit) (Bokemeyer et al., 1983), shown in Figure 5.1. The lower graph shows the positron energy spectrum for the angular region $11° \leqslant |\Delta\theta| = |\theta_1 - \theta_2| \leqslant 19°$ and $89.2° \leqslant \Sigma\theta = \theta_1 + \theta_2 \leqslant 89.8°$, where θ_1 and θ_2 are the scattering angles of the projectile ion and the target atom, respectively. Two peak-like structures are seen at positron energies of

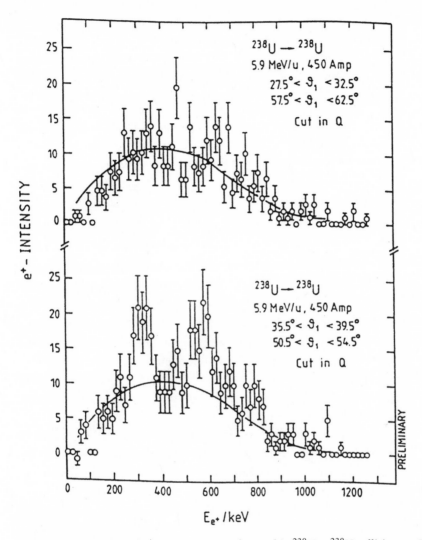

Figure 5.1. Two selected e^+-energy spectra observed in ^{238}U + ^{238}U collisions at 5.9 MeV/u. The solid lines represent the theoretical spectra normalized to the upper spectrum. Notice that the energy peaks appear only in the lower graph. From Bokemeyer et al. (1983).

approximately 320 keV and 590 keV. The peaks do not appear in the upper graph, which is the positron energy spectrum for $25° \leqslant |\Delta\theta| \leqslant 35°$ under otherwise identical conditions. The EPOS I experimenters were aware of the problems of both reproducibility and selectivity. In discussing the significance of their results, Bokemeyer and company (1983) stated:

However, before drawing any far reaching conclusions from the experimental data presented, we feel that the following questions should be solved. First of all one must show by further analysis that the structures are not produced by some yet unknown background effects *associated with one of the event selecting criteria.* *Secondly, the reproducibility of the effect has to be shown.* (p. 290, emphasis added)

The problem of reproducibility mentioned was not merely methodological and abstract. It appeared quite dramatically in the results presented by the ORANGE group (Kienle, 1983). They found peak-like structures appearing at positron energies of 370 keV and possibly at 720 keV and 950 keV. The structures appeared most strongly only for certain scattering angles, an effect similar to that seen by EPOS I. The energy of the lowest peak differs from that of EPOS I by approximately 50 keV and the 590 keV peak seen by the latter is not visible at all in the ORANGE results. The peaks at 720 keV and 950 keV observed by the ORANGE group were not seen by EPOS I; nor were they seen when the ORANGE group repeated their experiments.

An excess of positrons produced at low energy reported by the TORI group at the same time suffered from the same problem (Backe et al., 1983a). The question of the sensitivity of the observed results to the experimental conditions and what constitutes an adequate replication had appeared (Greenberg, 1983):

However, it is again disconcerting to find that the inconsistency observed for some of the measurements reported by Kienle is repeated here. A subsequent experiment was not able to reproduce the deviations from theory found in the initial data on the U + U system. This begins to suggest that controlling the bombarding conditions carefully may be a crucial ingredient in studying these effects. As I noted already, the underlying reason for this sensitivity is presently not understood, unfortunately. (p. 883)

Despite the problems he had noted, Greenberg's summary of the experimental situation at a 1981 conference was quite positive:

Thus, these last experiments, like the others we have discussed, suggest very convincingly that there are excess positrons above the dynamically induced background, but they go even further in pointing out that the additional positrons are associated with selected kinematic conditions possibly reflecting a focused nuclear reaction. (p. 886)

Unfortunately, theory did not provide either suggestions or justification for what those conditions and associated selection criteria should be, further complicating the problem.

Experimental work continued, but the uncertainty remained. For example, the EPOS I group reported a 316-keV positron line produced in U + Cm collisions (Greenberg and Greiner, 1982; Schweppe et al., 1983). Unfortunately, the group using the TORI spectrometer concluded that for the same reaction "no statistically significant structures have been observed in this experiment" (Backe et al., 1983b, p. 1840).

The existence of a peak near 300 keV was further supported by the ORANGE group (Clemente et al., 1984). However, the position and strength of this and other observed peaks are sensitive to the bombarding energy and the heavy-ion scattering angle, as shown in Figure 5.2. Further work by both EPOS I and ORANGE groups achieved, at least, internal consistency. The EPOS I group observed peaks in the Th + Th, Th + U, U + U, Th + Cm, and U + Cm systems (Cowan et al., 1985). Their average value for the peak energy was 336 keV. Similar consistency was found by the ORANGE group for the U + U and U + Th systems. Their peak energy was, however, approximately 280 keV. Under virtually identical conditions, different results had been obtained.

Both groups agreed that neither nuclear transitions nor spontaneous positron emission could be the cause of the positron lines. One could salvage spontaneous positron emission as a cause of the observed positron lines, but only by invoking changes in the charge configuration and the ionization states for the compound system that were regarded as physically unrealistic. Both groups suggested that the constant energy peak might have a common—and different—source: "*An obvious speculation is that the source of the monoenergetic positrons is the two-body decay of a previously undetected particle*" (Cowan et al., 1985, p. 1764, emphasis added).

Everyone—experimentalists and theorists alike—agreed that "a clear signal for a neutral particle could be provided by the detection of a

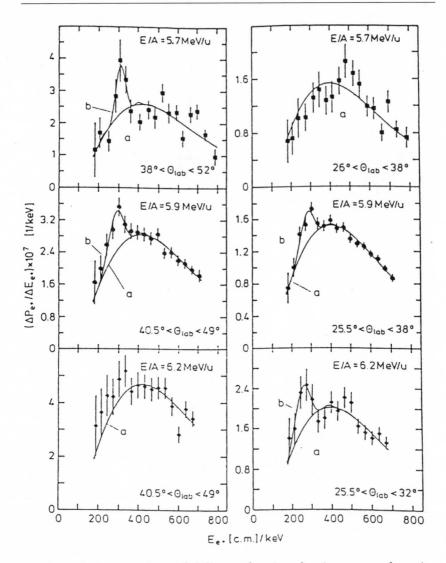

Figure 5.2. Positron creation probability as a function of positron energy for various bombarding energies and scattering angles. Notice that the peaks do not appear in all of the graphs, nor do they appear at the same energy in the graphs. From Clemente et al. (1984).

monoenergetic electron in coincidence with the peak positrons" (Cowan et al., 1985, p. 1764).

Electron-Positron States

The EPOS I group first reported a possible low-mass neutral particle (Cowan et al., 1986). This paper illustrated the dependence of the observed effects on selection criteria or cuts, discussed earlier in this chapter. In this experiment, the EPOS I spectrometer had been modified so that both electrons and positrons could be observed in coincidence and their respective energies measured. Striking effects appeared, but only under certain circumstances. Figure 5.3a shows the intensity distribution for all coincidence events as a function of the kinetic energies of

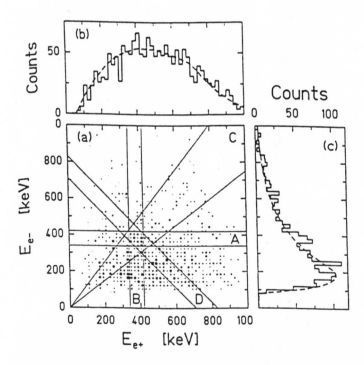

Figure 5.3. (a) Scatter plot of positron and electron energies; (b,c) the projections of the scatter plot. The projections along the energy axes are shown along with the wedge cut (C), $E_{e+} \approx E_{e-}$, and the cuts for constant electron (A), positron (B), and sum energy (D). From Cowan et al. (1986).

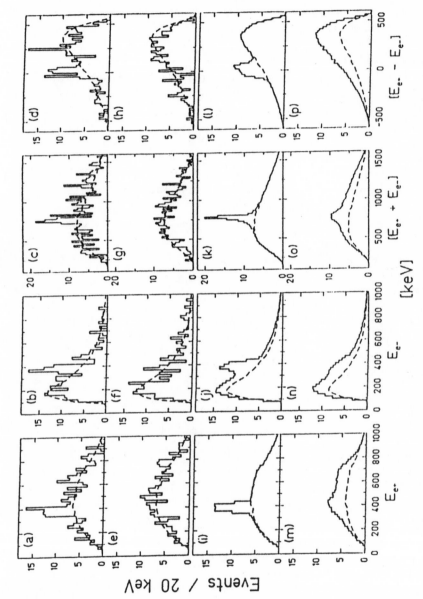

Figure 5.4. Projections of the measured intensity distribution onto (a,e,i,m) the E_{e+} axis; (b,f,j,n) the E_{e-} axis; (c,g,k,o) the $E_{e+} + E_{e-}$ axis; and (d,h,l,p) the $E_{e+} - E_{e-}$ axis. The columns correspond to the gates A–D of Figure 5.3. (i–l) and (m–p) are Monte Carlo calculations for two-body decay of a neutral particle and for internal pair conversion, respectively.

the positron and of the electron, E_{e+} and E_{e-}, respectively, for U + Th collisions at 5.83 MeV/u. Parts b and c of the figure show the projections of the distribution onto the E_{e+} and E_{e-} axes, respectively. No structure is apparent. The curves shown are the result of a Monte Carlo calculation that included both nuclear and atomic processes. It is a good fit to the spectra.

When cuts were made on the data, however, structures did appear. Figure 5.4a shows the positron energy spectrum obtained when the energy of the coincident electron was restricted to the range $340 < E_{e-} < 420$ keV. Figure 5.4b shows the complementary distribution for electron energy when the positron energy was restricted to the same region. Both graphs show peaks at approximately 380 keV, with widths of approximately 80 keV, indicating that a significant fraction of the coincident events had electrons and positrons with the same energy. The experimenters noted that the observed effect was six standard deviations above background.[4] (Equal energies are what one would expect if both particles resulted from the decay of a slowly moving particle.) This peak, although similar to those previously reported, is, in fact, different from them. EPOS I had previously found a positron peak at approximately 336 keV and the ORANGE group had found a peak at approximately 280 keV.

Figure 5.4c shows the number of events as a function of the sum of the electron and positron energies, $E_{e-} + E_{e+}$. The experimenters required $E_{e-} \approx E_{e+}$, which, when combined with the kinematic broadening of the energies, resulted in cut C (the wedge cut in Figure 5.3a). "The resulting sum-energy spectrum contains a narrow peak, at a mean energy of 760 ± 20 keV, with 35.3 ± 9.4 events in excess of the fitted continuous background" (Cowan et al., 1986, p. 446). Figure 5.4d is the intensity of events as a function of the energy difference $E_{e+} - E_{e-}$, requiring that $E_{e+} + E_{e-}$ be constant (cut D in Figure 5.3a). A peak is also seen at $E_{e+} - E_{e-} \approx 0$.

None of these structures appeared in any of the projections for the energy regions adjacent to cuts A–D on either side (Figure 5.4e–h). They were quite sensitive to the energy cut. The authors were able to fit a Monte Carlo calculation to the observed peaks by assuming that neutral particle with mass 1.8 MeV was produced in the collisions (Figure 5.4i–l). The EPOS I group had found that a significant fraction of the electrons and positrons from coincident electron-positron pairs had approximately the same energy and were given off at large angles (~180°) to one another. This was compatible with the view that they were the result of the decay of a low-mass particle, produced at very low velocity in the center-of-mass frame of reference. They concluded that "Features associated with the electron-positron decay of a slowly moving neutral particle appear to be reflected in the observations involving electron-positron coincidences" (Cowan et al., 1986, p. 447).

The EPOS I group presented new results in June 1986 (Cowan et al., 1987). This paper included a history of the positron lines to that time, details of the modified EPOS spectrometer, the results on the low-mass electron-positron states previously presented by the group, and a detailed discussion of new results. The sensitivity of the experimental results to experimental conditions and selection cuts was made even more apparent in this paper.

The most striking new result was evidence for two additional low-mass electron-positron states at sum energies $E_{e+} + E_{e-}$ of 620 keV and 810 keV. As was the case with the previously reported 760-keV state, the effects did not appear in all of the data: Observing them required selectivity. Figure 5.5 shows the new results. The original figure caption read, "Results of a preliminary analysis of U + Th collisions near 5.87 MeV/u (Feb 1986). ($E_{e+} + E_{e-}$) and ($E_{e+} - E_{e-}$) projections for two subsets of data gated on beam energy, heavy-ion scattering angle and e^+ or e^- TOF [time of flight] *chosen to enhance the prominent lines at ~810 keV and ~620 KeV, respectively*" (Cowan et al., 1987, p. 117, emphasis added). The data in Figure 5.5a,b (the 810-keV state) were obtained with beam energy 5.87–5.90 MeV/u and TOF difference between the heavy-ion signal and the electron and positron signals set for "prompt" events (prompt events are those expected from the decay of a single particle). The data in Figure 5.5c,d (the 620-keV state) had beam energy 5.85–5.90 MeV/u, with the electron TOF set for prompt electrons with the positron TOF delayed by an average of 3 ns. The sensitivity to cuts is clear: Only one state appears in each set of graphs.[5] The peaks were sensitive both to the bombarding energy and TOF.

The sensitivity of the observed effect to the TOF cut had been found empirically. The TOF cut was applied because another result had been seen earlier at one time delay but not at others. No theory at the time predicted any sensitivity to the TOF of the electrons and positrons. There were other such effects. Some effects were seen only in data taken with a "fresh" target. Figure 5.6b shows a positron line observed only with data taken in the first hour of heavy-ion irradiation of the target. Figure 5.6a, the total data sample for the run, shows no such effect.

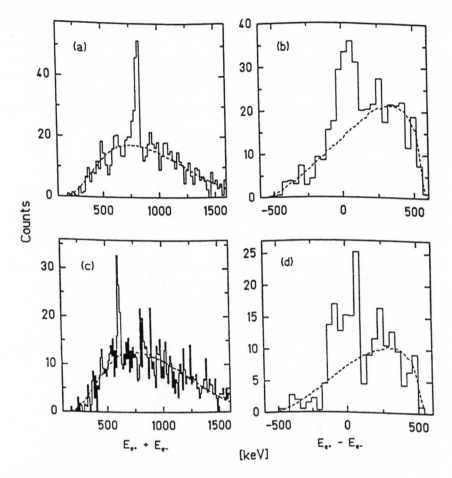

Figure 5.5. Sum- and difference-energy spectra for U + Th collisions, with cuts chosen to enhance the lines at (a,b) ~810 keV; (c,d) ~620 keV (see discussion in text). From Cowan et al. (1987).

This illustrates and emphasizes the problem of cuts. In studying a rare and previously unobserved phenomenon, one may very well have to apply cuts to see any effect at all. One might ask: Is the observed effect real or an artifact of the cuts? Is the experimenter tuning the cuts to create the effect? Is the sensitivity of the result to beam energy due to a real resonance phenomenon, as some physicists suggested, or is it only a statistical effect enhanced by the cut on beam energy? Similar questions arise for the TOF and target exposure cuts.

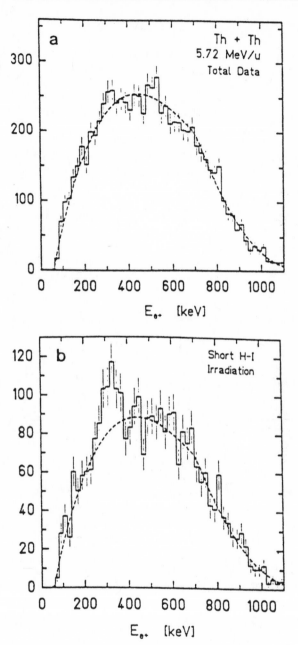

Figure 5.6. (a) Total positron energy distribution for Th + Th collisions at 5.75 MeV/u and 5.70 MeV/u. (b) Subset of the data selecting approximately the first hour of irradiation. From Cowan et al. (1987).

Although the experimental results were compatible with the production of a new particle, no theory or model could actually explain the results. Despite the theoretical difficulties, the experimental results on the positron lines and the possible low-mass electron-positron states were sufficiently credible that other experimenters searched for similar effects in other interactions in which such effects might be expected to appear. All but one of these searches was negative. No trace of a new particle was seen in radiative upsilon decay, electron bremsstrahlung, nuclear decays, muon or pion decay, or hadronic showers.

The only positive result reported in these searches was that of a suggestive peak in the sum-energy spectrum for electrons and positrons produced in e^+ + thorium collisions (Erb et al., 1986). The experimenters were quite positive about the previously reported positron lines:

The peaks observed in the energy spectra from heavy-ion collisions raise a variety of important questions. Established beyond any doubt by a series of heavy-ion experiments their properties are puzzling and their existence appears to be unexplainable in terms of conventional or atomic phenomena. (p. 52)

They concluded:

The width and energy of the peak structure in our data is similar to that found in the heavy-ion data [their sum-energy peak was at 670 keV, in contrast to the peaks found at 620 keV, 760 keV, and 810 KeV by the EPOS group], leading to the speculation that both may reflect a common underlying process. (p. 56)

The subsequent history of this particular effect is a microcosm of the entire episode. Attempted replications of the experiment gave positive, negative, and inconclusive results. There were, in addition, problems of interpretation. Peckhaus and collaborators (1987), for example, also searched for a peak structure in e^+ + Th collisions and found no effect. They noted, however, that the identification of the peak at 340 keV (the energy reported by Erb and collaborators [1986]) was complicated by the Compton edge in the scattering of annihilation γ rays. Bargholtz and collaborators (1987) found a small (2.5-standard-deviation) effect at the same energy, but remarked that the effects of multiple scattering and energy loss were large in their experimental apparatus and concluded: "The evidence for this peak is not considered conclusive" (p. L265). Wang and collaborators (1987) also looked for e^+e^- coincidences in the scatter-

Table 5.1.

Positron lines observed

E_{e+}^{cm} (keV)	FWHM (keV)	Collision System	Z_u	Bomb Energy (MeV/u)	Apparatus	Reference
238 ± 10	50	U + U	184	5.6, 5.9	ORANGE	Tsertos et al. (1987)
250 ± 5	34	U + Ta	165	5.9	ORANGE	Present work
261 ± 4	26	Pb + Pb	164	5.7	ORANGE	Present work
263 ± 5	24	U + Au	171	5.9	ORANGE	Present Work
Mean value 255 ± 7						
277 ± 6[a]	65	U + Th	182	5.9	ORANGE	Tsertos et al. (1985)
280 ± 6[a]	70	U + U	184	5.7, 6.2	ORANGE	Tsertos et al. (1985)
310 ± 5	~25[b]	U + Th	182	5.85–5.90	EPOS, e^+e^-	Cowan et al. (1986)
313 ± 10	~75	U + U	184	5.9	EPOS	Cowan et al. (1985)
316 ± 10	~75	U + Cm	188	6.05	EPOS	Cowan et al. (1985)
327 ± 10	~75	Th + Th	180	5.75	EPOS	Cowan et al. (1985)
330 ± 4	~13	U + Ta	165	5.9	ORANGE	Present work
334 ± 10	~75	U + Cm	188	6.07	EPOS	Cowan et al. (1985)
337 ± 4	33	U + Au	171	5.9	ORANGE	Present work
348 ± 10	~75	U + U	184	5.8	EPOS	Cowan et al. (1985)
349 ± 10	≥100	Th + U	182	5.82	EPOS	Cowan et al. (1985)
350 ± 5	39	Pb + Pb	164	5.7	ORANGE	Present work
352 ± 4	34	U + U	184	5.6, 5.9	ORANGE	Tsertos et al. (1987)
354 ± 10	~75	Th + Cm	186	6.02	EPOS	Cowan et al. (1985)
Mean value 337 ± 6						
375 ± 10	~75	Th + Ta	163	5.78	EPOS	GSI report
380 ± 5	~80[a]	U + Th	182	~5.86	EPOS, e^+e^-	Cowan et al. (1986)
380 ± 10	≥80	U + U	184	5.9	TORI	GSI report
405 ± 5	~40[b]	U + Th	182	5.87–5.90	EPOS, e^+e^-	EPOS report
409 ± 5	31	U + U	184	5.6, 5.9	ORANGE	Tsertos et al. (1987)
Mean value 396 ± 5						

Source: Koenig et al. (1987).

[a] These values are not included in the determination of the mean values.

[b] Widths observed for the e^+e^- sum energy peaks.

ing of positrons from thorium. They found no effect, but they did observe a Compton edge due to the scattering of annihilation γ rays.

However, Sakai et al. (1988) investigated positron scattering from Th, U, and Ta. They found evidence for a very narrow peak at ~330 keV with a width of 3.7 keV, but only for the Th and U targets. There was no explanation for this unusual and unexpected difference. Sakai also reported that new data showed the existence of an energy peak at 409 keV, corresponding the sum-energy peak at ~810 keV reported by the ORANGE group (Sakai, 1989). To say the least, the situation was unclear.

The uncertainty concerning both the effects and their explanation was increased when the ORANGE group continued its work on the single positron lines, extending it to the subcritical systems Pb + Pb, U + Ta, and U + Au, in which spontaneous positron creation was not expected. Nevertheless, they found two lines at energies of ~258 keV and ~340 keV. The peak energy of these lines was, once again, independent of Z_u, the total charge of the system. They found, however, that the cross section for positron line production had a strong Z_u dependence, proportional to $Z_u^{22 \pm 2}$. This was in disagreement with the results previously reported by the EPOS I group that both the production of positrons and the angular distribution were independent of the total charge of the system. There was, in addition, a factor of 10 difference in the production rates. The ORANGE group regarded the qualitative similarities as more significant than the quantitative disagreement (Koenig et al., 1987).

The ORANGE group provided a summary of all the positron lines that had been observed up to that time (Table 5.1). They arranged these lines into three groups with mean energies of 255 ± 7 keV, 337 ± 6 keV, and 393 ± 5 keV. Note, however, that the measured line widths varied from 24–50 keV, 13–100 keV, and 31–80 keV, for the three lines, respectively. It is not at all clear that the three groups actually define three distinct positron lines. One might remark that the grouping of the measurements seems somewhat arbitrary. An equally good case could be made for a continuous spectrum or other groupings (Figure 5.7). The experimental situation with respect to these lines remained unclear.

The experimental uncertainty in the positron lines and failure to detect a low-mass electron-positron state in other experimental systems—combined with the lack of any acceptable theoretical explanation of lines or state—led theorists to search for another interaction that might provide both an explanation of the effects and a possible experi-

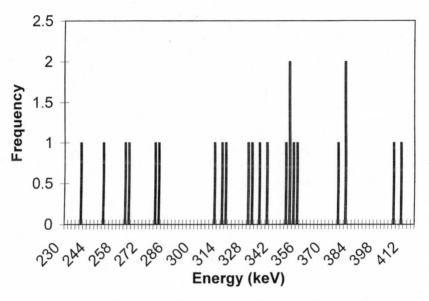

Figure 5.7. Energy of the reported positron lines. Data from Koenig et al. (1987).

mental confirmation. The suggested interaction was Bhabha scattering (scattering of electrons and positrons). Although some work on this problem at energies of a few million electron volts had already been done, it had been a neglected area of study. The problems of the GSI results provided a motivation for more detailed study.

Experimenters were quick to investigate Bhabha scattering in the appropriate energy region. Once again, experiment gave conflicting answers. Six of the searches were negative: No resonant structures were seen in the results reported by Mills and Levy (1987), Connell et al. (1988), Lorenz et al. (1988), Tsertos et al. (1988a,b), and Van Klinken et al. (1988). The failure to observe a predicted effect that should have been seen in a different system cast doubt on the correctness of the original results. Three experiments found positive, but small, effects. Maier and colleagues (1987) found their largest deviation from the fitted Bhabha spectrum of approximately 5% at an energy of 824 keV, in agreement with one of the lines found by EPOS I. This was confirmed in a later experiment by the same group, which found an average enhancement of 1.6% at an energy of 810 keV (Maier et al., 1988). Von Wimmersperg and collaborators (1987) found a deviation from the Bhabha fit of about 6%

at an energy of 710 keV. They noted, however, that this result (and that of Maier et al. [1987]) was not easily reconcilable with the EPOS I observation of low-mass electron-positron states, which was that approximately 50% of the events produced in the energy region where the peak occurred were in the peak.

A flurry of new experimental results on the possible low-mass states appeared in early 1989. These were presented at the Moriond Workshop that was held 21–28 January 1989.[6] Once again, the contradictory results added to the confusion. Kienle (1989) presented results obtained by the ORANGE group with a new experimental apparatus.[7] The apparatus consisted of two orange-type β-ray spectrometers placed back to back. One was set to detect electrons and the other positrons. In the total data set for electron-positron coincidences for U + U and U + Pb collisions at 5.9 MeV/u, "no prominent structures were observed in these spectra" (Kienle, 1989, p. 71). When the angle between the electron and the positron was restricted to 180 ± 20° and the difference in energy between the positron and the electron limited to -150 keV $\leq E_{e+} - E_{e-} \leq 0$ keV, structures were observed in the sum-energy spectra at energies of 540 ± 16, 640 ± 10, 716 ± 10, 809 ± 8, and 895 ± 10 keV (Figure 5.8). These features appeared in both the U + U and the U + Pb collisions. No structures were seen for events with $E_{e+} - E_{e-} > 0$. The experimenters (Kienle, 1989) concluded that:

The present results from the sum-energy coincidence spectra are consistent with the energies of e^+ lines which we presented before. . . . the ensemble of lines, available in a wide range of collision systems suggests a family of resonances with invariant masses of ~1.54, ~1.66, ~1.72, ~1.83 and ~1.93 MeV/c² respectively. Two of these lines (1.66 and 1.83 MeV/c²) may be identified with the sum peak at ~620 keV and ~810 keV found recently by the EPOS collaboration in U-Th collisions. *The line intensities in this work however are approximately by a factor of 10 lower compared to those of* (Cowan, Greenberg et al. 1987). *No indication of a sum line at 760 keV, as reported by the EPOS collaboration (Cowan et al. 1986) was found.* (p. 74, emphasis added)

Once again, the observation of an effect was strongly dependent on the cuts applied. In this case, there were plausible physical reasons underlying the cuts. In the new apparatus, the electrons were detected in the forward direction relative to the beam, whereas the positrons were detected in the backward direction. This should result in an observed energy dif-

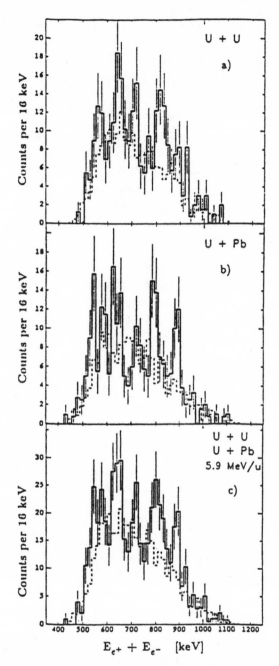

Figure 5.8. Sum-energy spectra from U + U and U + Pb collisions. From Kienle (1989).

ference due to the Doppler shift caused by the center-of-mass motion of the particle produced.[8] The electron should have a higher energy than the positron. (Note, however, that the original EPOS I result reported electrons and positrons with equal energies, the result one expected for the two-body decay of a very slowly moving particle.) In addition, the region of positive energy differences was where the background due to internal pair conversion, a nuclear process, was expected to be most pronounced. The angular restriction (180 ± 20°) was applied to select electrons and positrons resulting from the decay of a single particle. The decay particles should be emitted back-to-back (i.e., at 180°) in both the center-of-mass and the laboratory frame of reference because of the low center-of-mass velocity.

The EPOS I group also presented new results at this workshop (Bokemeyer et al., 1989). They found evidence in both U + Th and U + Ta collisions for resonances at 610, 750, and 810 keV. As before, the observed effects were dependent on the experimental conditions or the selection criteria. The resonances observed were quite sensitive to the bombarding energy and the TOF difference between the electron and the positron. For the U + Th collisions, for example, the 610-keV resonance is barely visible in Figure 5.9a3, but is quite pronounced in Figure 5.9b3. The difference between the two figures is that the bombarding energy for Figure 5.9a3 is 5.87–5.90 MeV/u, whereas for Figure 5.9b3 it is 5.86–5.90 MeV/u. Thus, increasing the energy range accepted by only 0.01 MeV/u changed the character of the observed peak considerably. The 810-keV peak seen in Figure 5.9a1 (3σ level) was enhanced to a 6σ effect by restricting the TOF difference to prompt events. The 610-keV peak, however, does not appear in the prompt events, but does appear in the remaining nonprompt events. Similar cut effects were seen in U + Ta collisions. In the U + Ta run, an additional feature had been added to help decide if the electrons and positrons were due to the decay of a single particle: Counters were added to identify into which hemisphere—forward (F) or backward (B)—the decay particle was emitted. For a two-body decay of a single particle, one expects the electrons and positrons to be emitted into different hemispheres (electron F and positron B, or vice versa). They found that only the 810-keV state was consistent with this condition. In fact, the 748-keV line seemed to be caused primarily by events in which both leptons were emitted into the forward hemisphere. The experimenters concluded:

The presence of narrow electron-positron sum-energy lines in heavy-ion collisions seems to be undebatable. The persistency with which these lines occur in independent experiments—together with their statistical relevance reaching values up to 6σ—makes their interpretation in terms of statistical fluctuations rather unprobable, if not impossible. The absence of the 760-keV line observed in the first coincidence experiments in ^{238}U + ^{232}Th collisions in our more recent runs (presumably scanning the beam energy region of our previous experiment), however, is not easy to understand. (p. 88)

Nevertheless, the EPOS results were inconsistent with the new ORANGE results and the previous results of the EPOS group.

Yet another experimental group reported results at the 1989 Moriond Workshop. The TORI group detected both electrons and positrons produced in collisions U + Th at 5.85 MeV/u (Rhein et al., 1989), almost identical conditions to those used by the EPOS I group in the experiment reporting their first positive result for a low-mass particle. The TORI apparatus could detect electrons in one of two conditions: the 0° condition ($0° \leq \theta_{e+e-} \leq 115°$) and the 180° condition ($67° \leq \theta_{e+e-} \leq 180°$). Using the 180° condition (that expected for the decay of a single particle) and applying the "wedge cut" ($E_{e+} - E_{e-} \approx 0$) used by EPOS I, they found no structure in their sum-energy spectrum. In contrast to the EPOS result, however, they did see structures in the 180° sum-energy spectra at 551, 642, and 749 keV, but only when the energy difference between the positron and the electron was not close to zero. In fact, the structures became more pronounced as the energy difference was increased.

Evidence against the existence of a low-mass particle in experiments on other interactions was also presented at Moriond. Van Klinken and collaborators (1989) summarized recent work on Bhabha scattering. He reported that five experiments—conducted at Giessen, Grenoble, München, Groningen, and Stuttgart—had all searched for a particle in the mass region around 1.8 MeV/c^2 and he concluded "that so far no resonances have been found within meaningful limits" (p. 147). The Grenoble group presented their own negative result (Schreckenbach, 1989). Negative searches in the reactions $e^+e^- \rightarrow n\gamma$ (Schreckenbach, 1989) and in nuclear decay (Sona et al., 1989) were also reported.

The final EPOS I results were published in 1990 (Salabura et al., 1990).[9] They were essentially the same as those presented at the 1989 Moriond Workshop:

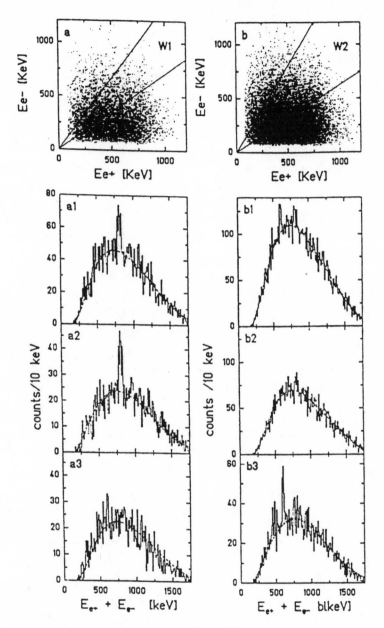

Figure 5.9. Sum-energy spectra in ^{238}U + ^{232}Th collisions. Beam energy is: (left) 5.87–5.90 MeV/u; (right) 5.86–5.90 MeV/u. The 810-keV state appears only for prompt events, whereas the 620-keV state appears only in nonprompt events. From Kienle (1989).

Three very narrow e^+e^- sum-energy peaks around 610, 750, and 810 keV have been observed in U + Th as well as in U + Ta collisions at beam energies around the Coulomb barrier. As no processes involving conventional atomic and nuclear physics were found to describe their origin, the data were in particular confronted with the hypothesis that the lines are due to the two-body decay of neutral objects in an e^+e^- pair. Although the 810 keV sum-energy line observed in U + Th is consistent with the prompt two-body e^+e^- decay of a neutral object if created nearly at rest in the heavy-ion center-of-mass system, the other lines require at least a considerably more complicated scenario if they are to be explained in the context of a two-body decay. (p. 153)

Somewhat later the ORANGE group presented their last results before an upgrading of both their apparatus and the accelerator (Koenig et al., 1993). They reported several lines in the energy range from ~550 keV to ~810 keV in the sum-energy spectra obtained from U + U, U + Pb, and U + Ta heavy-ion collisions. Not all of these lines were seen under all experimental conditions. As was the case for the EPOS I results, some of the lines were consistent with the two-body decay of a neutral object, whereas others were not.

Further evidence against the existence of such a neutral object was provided by continuing experiments (Wu et al., 1992) on Bhabha scattering:

Within statistical uncertainties (0.27%) no evidence has been observed for deviations from Bhabha scattering over the entire invariant-mass region 1560 keV/c^2 < M_{xo} < 1860 keV/c^2 that can be associated with the e^+e^- sum-peak energies in the GSI heavy-ion experiments. (p. 1729)

Let us summarize the situation as of 1993. Discrete energy lines had been reported in the positron energy spectra obtained from heavy-ion collisions (Table 5.1). These lines had various energies and did not appear in all of the systems studied, nor were they observed at all bombarding energies. They also seemed to be sensitive, at least on occasion, to such other experimental conditions as the scattering angle and the exposure time of the target. There were also problems concerning the reproducibility of the observations under seemingly identical experimental conditions.

Similarly, peaks had been observed in the sum-energy spectra of electrons and positrons produced in such heavy-ion collisions. These results exhibited not only the same kinds of sensitivity as did the single positron lines, but they also seemed sensitive to the TOF difference between the electron and the positron, their energy difference, and the angle between them. Some of these electron-positron peaks were consistent with the two-body decay of a neutral object, but others were not. No evidence of such a neutral object was seen in other interactions, particularly in the Bhabha scattering of electrons and positrons, in which one would also expect to observe effects if the sum-energy peaks were real.

Nevertheless the consensus seemed to be that something unusual and unexpected was being observed and that both the positron energy lines and the sum-energy peaks were real effects. Greiner and Reinhardt (1995), two theoretical physicists who had worked extensively in the field, summarized the experimental and theoretical situation:

During the last decade the development of this field [spontaneous positron emission] was overshadowed by the spectacular narrow lines in the positron spectrum and later the monoenergetic electron-positron pairs discovered by the EPOS and ORANGE groups at GSI. . . .

This apparent universality of the positron lines has created much excitement and led to the belief that some fundamental new process had been discovered. A large variety of speculations, most of them based on very shaky ground, were put forward. The most natural explanation for a constant line energy and two-body decay characteristics would be the creation and subsequent decay of a new elementary particle, e.g. the axion. This, however, soon could be ruled out by various arguments, in particular by many control experiments (high-energy beam dump searches, pair production in nuclear transitions). . . .

However, a probably fatal blow was dealt at the hypothesis of a new particle, be it elementary or composite, by a set of experiments looking for resonances in electron-positron scattering in the mass region around 1.8 MeV. The outcome of these experiments (which are sensitive to resonances with a width down to the μeV level and have fully covered the relevant region of lifetimes) has been completely negative.

Thus one has to conclude that the GSI positron lines are only observable in experiments which involve heavy ions. Unfortunately the experimental results have changed considerably over time and it is difficult to decide which of the data are to be considered reliable.[10] (pp. 217–18)

Experimentalists were more positive. A 1995 paper published jointly by the ORANGE and EPOS II groups (Bar et al., 1995) offered the following summary:

Narrow sum-energy lines have been identified in ^{238}U + ^{238}U, ^{238}U + ^{232}Th, ^{238}U + ^{208}Pb, ^{238}U + ^{206}Pb, and ^{238}U + ^{181}Ta collisions in EPOS and ORANGE experiments. The sum energies of the lines depend on the collision system and kinematical parameters. The lines group around ~550, ~620, ~740, and ~810 keV if the assumption of a common origin can be made. The observed cross sections vary from $d\sigma_{e+e-}/d\Omega_{ion}^{CM} \approx 0.1$ μb/sr (815 keV, ^{238}U + ^{208}Pb) to 3.6 μb/sr (748 keV, ^{238}U + ^{181}Ta). The statistical significance of the lines reaches up to 6.5 σ (634 keV), but in some cases is limited. The experimental knowledge on production and decay channels is poor. (p. 241)

The APEX group (ATLAS Positron Experiment), a new player in the field, agreed (Ahmad et al., 1995a). It is clear that, at the very least, the EPOS and ORANGE results were regarded as sufficiently credible to merit further investigation.

The Search Ends

The papers published in early 1995 jointly by the EPOS II and ORANGE groups (Bar et al., 1995) and by the APEX collaboration (Ahmad et al., 1995a) marked the beginning of the final act of the drama and gave hints of what was to come. The joint EPOS II-ORANGE paper described the improvements that had been made in both experimental apparatuses. In particular, EPOS II was now able to detect both positrons and electrons in either side of their detector, increasing the yield of events. The groups reported results for an experimental run on the reaction ^{238}U + ^{181}Ta at heavy-ion bombarding energies from 5.98 to 6.07 MeV/u. EPOS I had previously reported a sum-energy peak at ~748 keV in this reaction. The EPOS II group (Bar et al., 1995) reported a similar peak in their new run:

A sum-energy line around 740 keV is identified in the first run, with an energy uncertainty of ~10 keV (fig. [5.10]). The line is poorly visible on the total spectrum of 45000 pair events which can be fully described otherwise by a MC [Monte Carlo]-calculation based on quasi-atomic and nuclear pair production, reproducing all global dependences established by the previous experiments. (p. 242)

Making cuts on the data requiring an interaction distance larger than 18.6 fm (this is related to the heavy-ion scattering angle), and a positive energy difference between the positron and the electron enhanced that

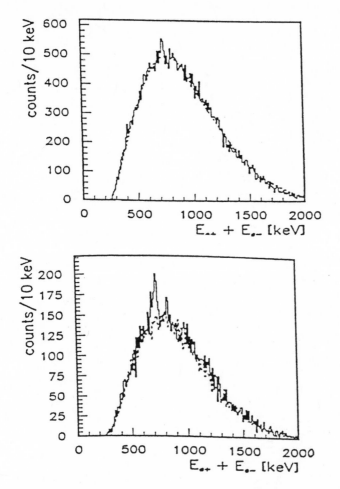

Figure 5.10. Sum-energy spectra for U + Ta collisions at 5.98 and 6.07 MeV/u. (top) Total spectrum; (bottom) spectrum selected for scattering angle. From Bar et al. (1995).

peak to a 5.5σ effect (Figure 5.10, bottom graph). In a foreshadowing of later events, however, the peak did not appear in a subsequent run under seemingly identical circumstances:

In the second run with ~3 times more total pair events the existence of a comparable line is not evident. Changes in the experimental set-up are presently [being] investigated to clarify if these could influence the observation of the line. Nevertheless, this apparent inability to properly set the experimental conditions

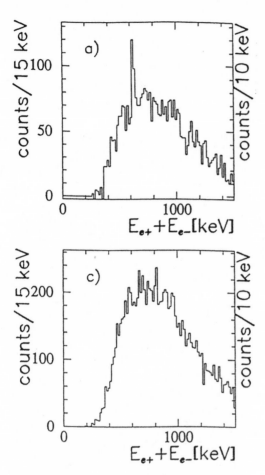

Figure 5.11. Sum-energy spectra obtained by EPOS II. The peak that appears in (top) the early data disappeared in (bottom) the larger, total data sample. From Ganz (1995).

again supports the assumption that important parameters for the source of the lines still remain unidentified. (p. 242)

(A similar effect is shown in Figure 5.11. A peak that is visible in an early run is not seen in a later run that had more statistics and was taken under identical conditions and used identical cuts.) The groups did not, at this time, interpret their failure to reproduce the observed effect as casting doubt on the reality of that effect, or as evidence that the observed effect might be an artifact created by the cuts, but rather as demonstrating their lack of knowledge of all of the important experimental conditions and their inability to reproduce them.

The APEX group, in a paper contiguous to the EPOS-ORANGE paper (Ahmad et al., 1995a), reported preliminary results. They also examined the reaction $^{238}U + {}^{181}Ta$ at energies comparable to those used by EPOS II. They found no peaks in either the positron energy spectrum or in the sum-energy spectrum. They investigated the positron-electron sum-energy spectra in the $^{238}U + {}^{181}Ta$ system (at bombarding energies of 5.95, 6.10, and 6.30 MeV/u) and the $^{238}U + {}^{232}Th$ system (at 5.95 MeV/u). These were the systems and the energies at which both the ORANGE and the EPOS I groups had previously reported peaks. APEX found "no statistically significant evidence for sharp sum-energy lines."

The group applied selection criteria in an attempt to observe the sum-energy peaks previously reported at 760 and 809 keV in the $^{238}U + {}^{232}Th$ system. These peaks had also been reported to be consistent with the two-body decay of a neutral object. Their results are shown in Figure 5.12 (Ahmad et al., 1995b). No hint of any peak is seen. The events in the upper curve of Figure 5.12a are those selected by the "wedge cut" (approximately equal electron and positron energies) previously used by EPOS I and ORANGE. The histogram is the spectrum of uncorrelated pairs generated by summing the energies of electrons and positrons from different events (event mixing):

The dashed peak, superimposed on the event-mixed spectrum, corresponds to the signal expected from the decay of an isolated neutral object of mass 1.8 MeV/c^2, produced with the cross section given in [Salabura et al. (1990)] $(d\sigma/d\Omega_{HI} \sim 5 \mu b/sr$—the pair production cross section averaged over the heavy-ion detector acceptance). (p. 2660)

Further cuts on the solenoid azimuthal angle and energy correlations expected for two-body decay are shown in the lower curve of Figure 5.12a. Once again, the effect expected on the basis of the previous results was not seen. Nor was the 748-keV peak previously reported in the U + Ta system observed.

The absence of the reported sum-energy lines in our data is puzzling. The origin of this apparent discrepancy may lie in so far unknown characteristics of the phenomenon. The overlap between the acceptance of APEX and that of the previous experiments is large. Nevertheless it is conceivable that the energy and angle correlations of the lepton pairs are such that they escape detection

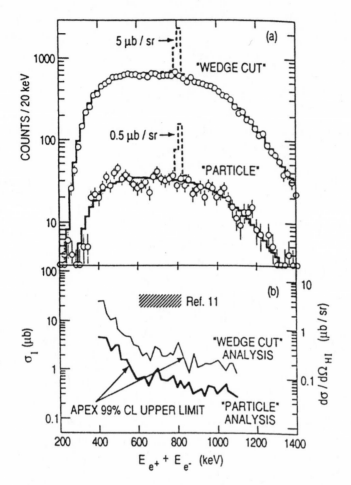

Figure 5.12. Sum-energy spectra for ^{238}U + ^{232}Th at 5.95 MeV/u, analyzed according to the expectations for the isotropic decay of a particle produced at rest in the center of mass. The "particle analysis" was for events with positrons and electrons with an opening angle of approximately 180°. From Ahmad et al. (1995b).

in our apparatus, although rather extreme situations are required for this to occur. (p. 2661)

The APEX group examined other possibilities for explaining the discrepancy and concluded, "Nevertheless, we believe that the results of the present experiment represent a real disagreement with the previous observation" (Ahmad et al., 1995b, p. 2661).

Strong evidence against the existence of the sum-energy peaks was also provided by the EPOS II group: "special care was taken to subtend as closely as possible the beam energy range investigated by EPOS I" (Ganz et al., 1996, p. 6). In this run, EPOS II obtained far more data than had been acquired by EPOS I; 10 times more for the U + Th system and 25 times more for the U + Ta system. In analyzing their data, they used cuts identical to those used by EPOS I in both the beam energy and the wedge cut, (i.e., the energy sharing between the electron and the positron). They omitted the TOF cut because that cut would have complicated the comparison between the old and the new results. They noted that the EPOS I peaks had been visible using only the two cuts that EPOS II was using. The EPOS II results for the three most prominent peaks observed by EPOS I— the 608- and 809-keV lines in the U + Th system and the 748-keV line in the U + Ta system—are shown in Figure 5.13. "No evidence for narrow line structures at sum energies given by the EPOS I experiment has been observed" (Ganz et al., 1996, p. 7). "As indicated in the right column of Fig. [5.13], where the differences between the measured sum-energy spectra and the normalized event-mixing distribution are displayed, the expected line yields are clearly in contradiction with the new high-statistics data" (p. 9). They concluded, "In summary, all our attempts led to the same result. We did not succeed in reproducing with our statistically improved data basis the e^+e^- sum-energy lines reported in EPOS I" (p. 10).

The difference between the EPOS I and II results was clearly troubling and the experimenters discussed various possibilities for explaining the discrepancy. They noted that the APEX results were consistent with those of EPOS II, and that they covered an angular range for electron-positron emission that was forbidden in EPOS II, eliminating that as a possible explanation of the discrepancy. They also concluded that differences in acceptance, unknown experimental parameters, or background processes could not explain the discrepancy. Still the question remained: Why had EPOS I observed peaks in the sum-energy spectrum?

The EPOS II group noted that judging the statistical significance of peaks enhanced by selection criteria, or cuts, was an extremely difficult problem. In Ganz and collaborators (1996), they remarked on the sum-energy line that they had reported earlier in Bar and collaborators (1995):

The 723 keV line [the energy calibration had changed between the two experiments] was only weakly visible in the total e^+e^- sum-energy spectrum but clearly

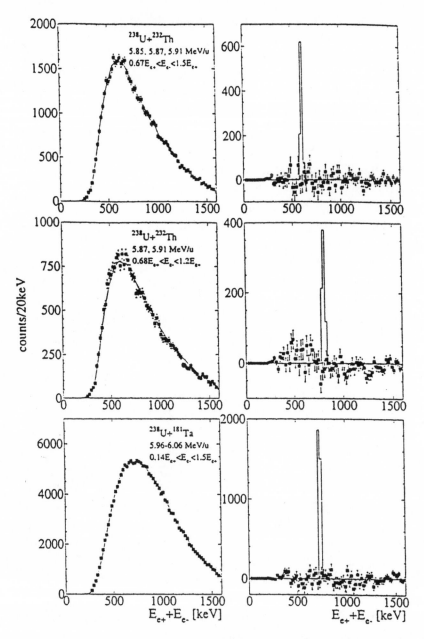

Figure 5.13. Sum-energy spectra obtained by EPOS II. The lines in the right panel are the yields expected for the 608-, 748-, and 809-keV states found previously by EPOS I. From Ganz et al. (1996).

seen (with a nominal significance of 5σ) when requiring cuts with respect to E_Δ [the energy difference between the electron and the positron] and the scattering angles of the heavy ions. However, not only was the energy of the line at variance with the previous value of 748 keV, also two follow-up, high-statistic experiments performed with an unchanged EPOS II set-up failed to reproduce the line [A similar effect for the line at 620 keV is shown in Figure 5.11]. (pp. 10–11)

The group further investigated the possibility that the effect was due to the cuts, suggested by the disappearance of the line with higher statistics:

The second example comes from an investigation suggested in Ref. 17 [Roe, 1992] where we took advantage of the enlarged data basis collected in the EPOS II experiment. We randomly distributed—on an event-by-event basis—the e^+e^- pairs collected at a certain beam energy into two subsets. While one of these sub-sets was kept as a reference sample, we searched for narrow line structure in the other subset by choosing different E_Δ and time-of-flight cuts. Surprisingly enough, we were able to find a cut—leading to a spectrum of similar statistics as in a typical EPOS I experiment—which enhances a 2σ-structure visible at 655 keV in the initial subset to a line of ≈ 5σ, which is comparable in width and intensity to those observed in EPOS I [see Figure 5.14].[11] However, applying the identical cut to the reference sample does not show any line structure at this energy [Figure 5.15]. (p. 12)

They concluded:

Both examples [the disappearance of a peak with higher statistics and the division of the data set into two subsets] underline that the statistical significance of spectra obtained by introducing selection criteria, which are acceptable when looking for something unexpected but which cannot be supported later by a coherent physical picture, has to be taken with great precautions. In this situation an independent reproduction based on a considerably larger data set is the only way to confirm the existence of a physical effect. Since we failed to demonstrate the reproducibility of the lines observed by EPOS I and derived cross-section limits which are a factor of up to 10 smaller than the values implied by the previous results, and in view of the negative results obtained by the APEX Collaboration the physical relevance of the EPOS I lines is questionable. (p. 12, emphasis added)

The ORANGE group also repudiated their earlier results (Leinberger et al., 1997). Their new experiment was designed to look for the ~635-keV

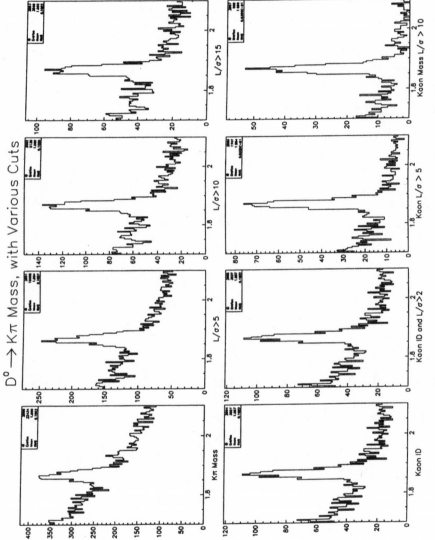

Figure 5.14. Graphs of the D^0 mass obtained for various cuts. Courtesy of Brian O'Reilly.

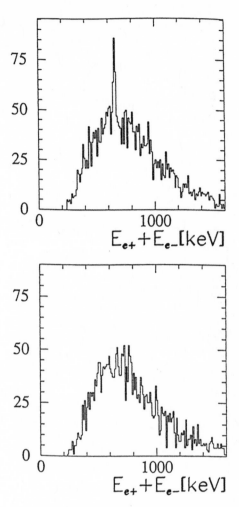

Figure 5.15. (top) The sum-energy spectrum obtained by tuning the energy-difference and TOF cuts to maximize the peak. (bottom) The effect of applying the identical cuts to the other half of the data: No peak is visible. From Ganz et al. (1996).

peak that was their most statistically significant result (6.5σ): "At improved statistical accuracy [by more than a factor of 10], the line couldn't be found in the new data" (p. 16). (See Figure 5.16.) The experimental conditions and the cuts used in the analysis of the data were extremely close to those used previously. The group also eliminated other possible differences in experimental conditions, such as target deterioration, as possible causes for the difference between the old and new results. "Taking into account that we have not found any evidence that the reported line [the earlier result] might be due to trivial effects or back-

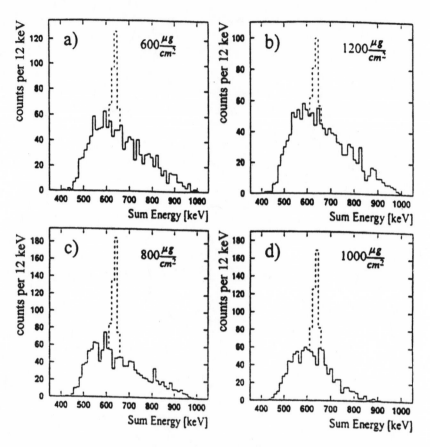

Figure 5.16. Sum-energy spectra from U + Ta collisions found by the ORANGE group. The superimposed peaks are those expected for the 635-keV state they had reported previously. From Leinberger et al. (1997).

ground processes, its statistical significance has to be reconsidered" (p. 21). Analysis of their new data had shown that the calculated background used in their earlier experiment was incorrect. Using the new background calculation the previously reported effect was reduced from 6.5σ to $\leq 3.4\sigma$. The ORANGE group also found, in agreement with EPOS II, that manipulating the cuts could produce electron-positron lines that disappeared with improved statistics.

Even before the publication of the new ORANGE results, the APEX and EPOS II experiments had provided the evidence that convinced virtually everyone working in the field that the sum-energy peaks previously reported by EPOS I, ORANGE, and others did not exist. In June 1996, a meeting was held at Oxford attended by physicists from the EPOS II, ORANGE, and APEX groups. At the end of that meeting, the conclusion was that the search was over and that no further experiments were needed.[12]

Discussion

Physicists, particularly those working on heavy-ion collisions, have concluded that both the positron lines and the sum-energy peaks observed by EPOS I, ORANGE, and others are not real effects. There is a strong suspicion that they are artifacts created by tuning the selection criteria applied to the data and by the effects of limited statistics. Taubes (1997), in fact, has suggested that such tuning even occurred in the acquisition of data in the early experiments:

Take what the EPOS physicists referred to as the top-hat criterion. Bokemeyer says that the EPOS physicists had noticed that what turned out to be peaks in the final analysis would first appear online as a top-hat shaped bulge in an otherwise smooth spectrum. So the experimenters would start collecting data at a particular energy or with a particular target, and if the spectra were smooth and flat, they would stop the experiment. "We would change the energy or target and try again," says Bokemeyer. "When the spectra started to look like a top-hat, this seemed to be the correct [conditions], and we would continue running without interruption." (p. 151)

As we have seen in the cases presented in previous chapters, experimenters do not use all of their data in producing a result. Cuts are always applied. This is not an unreasonable procedure. When one is looking for a small effect against a much larger background, cuts are needed to enhance the signal. (Recall the K_{e2}^{+} branching ratio experiment, discussed in Chapter 1.) One might, however, legitimately worry that the cuts are being tuned to enhance the effect. We have seen that it is possible to create a peak by tuning the cuts and we have also seen the safeguards taken to guard against this. The problem becomes even more difficult when (as is true in this episode) the result may depend on a large number of parameters, each of which may be used for selection, and the theory provides no guidance as to what the important parameters might be. The evaluation of the results was also made more difficult by their apparent, albeit not exact, replication. The question that must be answered, however, is whether the result is real or is an artifact created by the cuts.

The physicists involved in this episode were quite aware of this problem from the very beginning. Recall that Bokemeyer had urged caution in interpreting the first report of positron lines (see page 98). The many repetitions of the experiments under both very similar and different conditions were attempts to establish the correctness of the results by showing that they were reproducible; and were also efforts to acquire a physical understanding of the systems involved. Unfortunately, rather than clarifying the situation, the repetitions made it more complex. One might speculate that early in the investigation, the criteria for reproducibility were more lenient than they were later. Rough agreement was "good enough" to encourage pursuit,[13] or further investigation of a phenomenon: It might not be sufficient to establish the reality of an effect.

The discord was resolved when two experiments with higher statistics and therefore more evidential weight (EPOS II and APEX) found no evidence for the previously reported results. (This negative result was supported by subsequent ORANGE results.) In addition, EPOS II demonstrated that a peak found in a limited subset of their data, comparable to that of the earlier experiments, disappeared when the full data set was analyzed. They also showed that by choosing suitable cuts, they could create a peak similar to those found previously. That the effect disappeared when identical cuts were applied to the other half of the data suggested that the peak was, in fact, an artifact created by the cuts.[14] This also suggested that the limited-statistics peaks reported earlier might also be artifacts. The EPOS II group had shown that they would have detected a peak had one been present and that they could artificially create such a peak. As Dirk Schwalm, co-spokesperson for EPOS II, remarked, "I think we all overestimated the statistical relevance of the peaks we saw. It sounds a bit silly in the end, 10 years later, but I think that's what happened" (quoted in Taubes [1997], p. 151). The fact that the effects did not also appear in other interactions, such as Bhabha scattering, in which one would have expected them had the sum-energy peaks been real, provided further evidence against the reality of the peaks.

It was ultimately decided that the results were wrong. The decision that there were no low-mass electron-positron states was arrived at, as I have shown, on the basis of experimental evidence and rational discussion and criticism.

Death by a Thousand Cuts: Some Conclusions

In the first five chapters of Part I, we have seen several types of selection criteria that have been applied to data and analysis procedures. The cuts have ranged from Millikan's legitimate exclusion of data obtained when he was not sure that his experimental apparatus was working properly[15] to the very complex tuning of analysis cuts that produced an artifact in the case of the suggested low-mass electron-positron states. It is clear that there is no single solution to the problem of whether an experimental result is an artifact created by the cuts. What may work in one case may not work in another. There are, however, some general strategies.

Consider, for example, robustness. This is an important method of demonstrating the validity of an experimental result and dealing with the problem of cuts. It was, in fact, used in each of the episodes discussed in Part I of this book. In the experiment to measure the K_{e2}^+ branching ratio (Chapter 1), for example, the experimenters varied both the range cut and the track-matching criterion over reasonable intervals and showed that the branching ratio found was robust under those variations. In the case of both gravity waves and the 17-keV neutrino (Chapters 2 and 4), robustness was again important. In the gravity-wave episode, Weber's critics used both their own preferred analysis algorithm as well as Weber's nonlinear algorithm and showed that they still found no gravity-wave signal. This was one of the strong arguments in favor of the critics' results and against the correctness of Weber's result, which appeared only when his algorithm was used. Similarly, in the case of the 17-keV neutrino, several experimenters used both a wide and a narrow energy range in their analysis and demonstrated that their conclusions did not change. In the decisive experiments that showed that the 17-keV neutrino did not exist, the experimenters demonstrated that the choice of analysis procedure was not a problem in their experiments. We also saw that Simpson's apparent failure to use robustness as a criterion led to his interpretation of an artifact of data analysis as a real effect in his reanalysis of Ohi's data. In the case of Millikan's measurement of the charge of the electron (Chapter 3), robustness was provided by subsequent measurements of that charge.

Robustness did not, however, provide an unambiguous solution to the problem in the episode of the low-mass electron-positron states. This was because the results obtained, although similar, seemed to be extremely sensitive to the experimental conditions, such as time of flight, bombarding energy, scattering angle, and the equality of electron and positron energies (the wedge cut). Varying these conditions seemed to make the effects vary or disappear: The results lacked robustness. Were the variations a real sensitivity to the conditions or were they artifacts? There are, after all, many phenomena studied in science that exhibit such sensitivity.[16] But in this episode, more careful analysis subsequently showed that tuning the cuts could produce such results.

How similar must two experiments be to count as replications and how similar their experimental results to count as confirmation? How similar the conditions or effects must be can be decided only on a case-by-case basis. In the episode discussed in this chapter, the sensitivity to experimental conditions was shown to be an artifact. In the case of the J/Ψ, as discussed in note 16, the sensitivity of the result to the experimental conditions led to an important discovery.

Showing that cuts can create the observed effects also played a significant role in these episodes. Thus, Kafka, analyzing his own data and varying his threshold criterion showed that he could create an apparent gravity-wave signal. The same effect was shown by Levine and Garwin using a computer simulation. In the episode of the 17-keV neutrino, Bonvicini demonstrated, also using a Monte Carlo calculation, that analysis cuts combined with limited statistics could produce effects that might mask or mimic the presence of the proposed particle. Conversely, arguing that the applied cuts cannot create the observed effect increases confidence in the result (as was the case for the K_{e2}^+ branching ratio experiment). It should be emphasized, however, that demonstrating that an effect can be produced by applying selection criteria can only cast doubt on an experimental result. It cannot demonstrate that the result is incorrect. In the case of both gravity waves and the low-mass electron-positron states, other arguments were both needed and provided.

Sometimes one can argue that an experimental result is not an artifact by the use of a surrogate signal. If the apparatus can detect such a signal, then it argues that the experimental apparatus and the analysis procedure are working properly. This was the case in the episodes of gravity waves and the 17-keV neutrino. Weber's critics were able to detect a pulse of acoustic energy injected into the antenna that mimicked the effect expected for gravity waves. The Argonne group was able to detect the kink created by the composite spectrum of ^{35}S and ^{14}C, which served as a surrogate for the effect expected for the 17-keV neutrino. Such a procedure tests the proper operation of the experimental apparatus and the analysis procedure, including the cuts.

Should the lack of a universal procedure to guard against results that are artifacts of the selection criteria cause us to doubt both experimental results and the science based on those results? I think not. Difficulty is not impossibility. Although, as we have seen, the validity of results may be difficult to establish, it is not impossible to do so. In each of the episodes presented in Chapters 1–5, the question of whether the result was an artifact was answered. It would be an error to conclude that because three out of the five cases discussed had results that were artifacts of the selection criteria that this is typical of experimental results in physics. The episodes were chosen precisely because there were discordant results and the selection criteria were important.[17] The K_{e2}^{+} branching ratio experiment is the norm, not the exception. Cuts may be ubiquitous, but they need not be fatal.

6

"Blind" Analysis

As we have discussed in previous chapters, there is at least a strong suggestion that experimenter bias was a factor in Millikan's measurement of the charge of the electron. Millikan excluded data and engaged in selective calculational procedures to produce his desired result.[1] Similarly, Joseph Weber may very well have used selective analysis procedures in producing his claim that he had observed gravity waves. Selectivity was also an important issue in the episodes of the claimed existence of both the 17-keV neutrino and of low-mass electron-positron states.

Recently, considerable attention has been devoted to trying to eliminate such experimenter bias by using a procedure known as blind analysis, in which "the physics result is kept hidden until the analysis is essentially complete" (Burchat et al., 2000, p. 3). This is not the first time such a procedure has been used.

In 1964, Murray Gell-Mann and George Zweig independently proposed the quark model of interacting hadrons, the strongly interacting particles (Gell-Mann 1964; Zweig 1964). One consequence of this empirically successful theory was that the fundamental constituents of the hadrons—quarks—would be fractionally charged, that is, having charge $\pm 1/3e$ or $\pm 2/3e$, where e is the charge of the electron. Experimental searches for these fractionally charged quarks were conducted, and by the early 1970s, a consensus had been reached that they had not been observed (Jones, 1977). That led to the idea of quark confinement (i.e., the force between the quarks increased with distance, so that free quarks could never be observed).

In the early 1980s, William Fairbank and his collaborators claimed that they had, in fact, observed fractional charges of $\pm 1/3e$. This result was an anomaly for the quark confinement theory. The experiment involved levitating superconducting niobium spheres in a magnetic field, a modern version of the Millikan oil-drop experiment. Their results were quite consistent, arguing for the correctness of the result.[2] "Out of 26 repeat measurements, we have observed 11 residual charges, in every case of $\pm 1/3e$" (LaRue et al., 1981, p. 967). The experimenters also noted that the residual charges they observed "fall into three groups which have weighted averages of $(-0.343 \pm 0.011)e$, $(0.001 \pm 0.033)e$, and $(+0.328 \pm 0.007)e$" (p. 967). That only residual charge values of 0 or $\pm 1/3$ were observed supported their claim that fractional charges, and thus free quarks, existed. This was in disagreement with both a highly confirmed theory and with other experimental results.

One problem with the Fairbank experiment was possible experimenter bias. The results of each measurement of residual charge were known to the experimenters when the final data selection was made. To guard against possible bias, Luis Alvarez suggested that a random number, unknown to the selector, be added to each residual charge result and subtracted only after final event selection had been made. This was done, and the results of the blind test were $(+0.189 \pm 0.02)e$ and $(+0.253 \pm 0.02)e$ (Phillips 1980). The observation of charges that were not 0 or $\pm 1/3e$ cast doubt on Fairbank's result. The application of blind analysis had been successful. It cast doubt on Fairbank's result and suggested that experimenter bias may have played a role in the production of that result.[3] To this day there is no credible evidence for fractional charges.

The Methods of "Blind" Analysis

The reasons for using blind analysis, along with possible problems due to experimenter bias are clearly stated in "Draft Guidelines for Blind Analysis in BABAR" (Burchat et al., 2000). This is an elementary-particle experiment that includes searches for rare decays, precision measurements, and time-dependent asymmetries in the decays of the B (primarily) and D mesons:[4]

The major motivation for a blind analysis is to adopt a technique which removes or minimizes Experimenter's Bias; the unconscious biasing of a measurement toward prior results or theoretical predictions. . . .[5]

There are a number of ways in which Experimenter's Bias can infect a measurement which can be eliminated with a blind analysis. First, the point at which the decision is made to stop working and present one's result can be influenced by the value of the result itself, and how it compares with prior results or predictions.[6] In a blind analysis the decision to stop and publish is made based on external checks, and not on the numerical value of the result. After all there is no information about the correctness of a measurement in the numerical value obtained; a blind analysis enforces this separation. Second, choices about the data to include, or the cuts to use, can be subtly biased, if the effect these choices have on the result is known. [As we have seen, the bias is not always subtle.] Often changes in an analysis, which change the data set, can affect the value of a result on a statistically reasonable way. A blind analysis ensures that such choices affecting the data sample do not bias the result. Third, the values and types of cuts to use can be biased by knowledge of the effect of these cuts on particular events in the data. In particular, for rare decay searches or measurements involving small samples a blind analysis removes the possibility that cuts are chosen to include or exclude particular events in the data. In this case a blind analysis ensures a statistically meaningful result. (p. 3)

The experimenters remarked that their use of blind analysis did not imply that nonblind analyses are flawed, but only that because it was possible to eliminate or reduce the possible effects of experimenter bias at little cost, they would employ that technique. Although the BABAR experimenters suggested different methods of blind analysis for each of their proposed measurements, I will discuss only the more general procedures.

For the search for rare decay modes (branching ratio $<10^{-4}$), the "hidden signal box" method was suggested. "In the hidden signal box method, we define a signal region in one or two variables in which the signal is expected to be concentrated. *Remaining blind in such cases means not looking at the events in the signal region in any way*" (Burchat et al., 2000, pp. 4–5). This prohibition held only until the background estimation was made. The group considered those rare decays in which a complete reconstruction of the exclusive final state could be made. In such an analysis, there is usually a pair of kinematic variables, whose range is restricted for the signal, but whose distributions are smooth and slowly

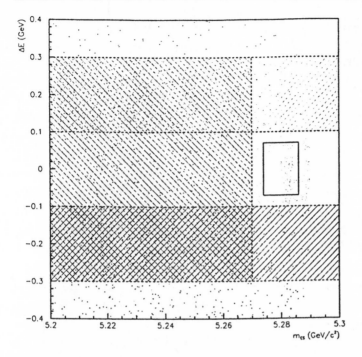

Figure 6.1. The (m_{ES}, ΔE) plane for rare-decay analysis. The shaded rectangle defined by 5.27 < m_{ES} < 5.29 GeV, $|\Delta E|$ < 0.1 GeV represents the blinded area, and the rectangle inside it, the boundary of the signal area. From Burchat et al. (2000).

varying for the background and at most weakly correlated. For the decays to be studied, the variables chosen were m_{ES}, the reconstructed mass of the B meson, and ΔE, the difference between the total measured energy and the beam energy. In this case a conservative signal box in the two variables was chosen, and an even larger blinding region (in which events were not examined until the backgrounds were determined) was established (Figure 6.1).

Data were taken in two different types of experimental runs: off-resonance, in which the energies of the colliding electron and positron beams were too low to produce B (or D) mesons, and on-resonance, in which the beam energies were exactly those needed to produce the desired particle.[7] For on-resonance data, the background was estimated in the following way. The distribution in one variable, let us say ΔE, was measured using data in which a cut in the orthogonal variable (m_{ES}) was made to avoid the signal region. Thus the entire region to the left of m_{ES} = 5.27 GeV/c^2 was used. Similarly, an m_{ES} distribution was found for the two strips outside the region $|\Delta E| < 0.1$ GeV. The background was then estimated by extrapolating the two distributions into the signal region. For off-resonance data, the entire plane could be examined and the background in the signal region counted. No real signal events were expected in these data. The estimation of the background using the two methods should agree within the experimental uncertainty. A second check can be provided by a Monte Carlo simulation of the off-resonance continuum. After the background estimation was made, the signal area would be examined.

For precision measurements of quantities such as Δm_d (related to the frequency of B^0–B̄^{0} oscillations[8]) or measurements of the lifetimes of the B and D mesons, the hidden offset method (discussed in the next paragraph) can be used. The experimenters noted that for each of these quantities, there existed very accurate prior measurements. Blind analysis was used to avoid any bias toward getting a result in agreement with those prior measurements. An interesting illustration of this method will be discussed later in this chapter.

The hidden offset method involves adding a fixed, hidden offset to the measured parameter in one of two ways:

$$x^* = \begin{cases} (1) & x + O \\ & \text{or} \\ (2) & 2<x> - x + O, \end{cases}$$

where O is the hidden offset, x is the measured parameter, and $<x>$ is either the previously measured value or the value contained in "Review of Particle Physics," the standard reference for the properties of elementary particles, compiled by the Particle Data Group (e.g., Groom et al., 2000). The analysis was then performed on x* rather than x, so that the value of the parameter found is unknown to the experimenters. The experimenters then concluded that "The hidden offset method is certainly not the only possible technique for precision measurements, *and there is no substitute for careful thinking about each particular analysis.* Finally, however, it is recommended that some blind analysis technique be adopted for each precision measurement" (Burchat et al., 2000, p. 7). They also suggested that care be exercised in choosing the analysis procedures when the data must, in fact, be studied directly because they do not fit easily into blind analysis techniques. The procedures suggested included avoiding examination of the final result until absolutely necessary, using a subsample of the data first to set procedures and cuts, not keeping track of the answer, and using optimization procedures and standard cuts where possible. Alan Schwartz (one of the members of the E791 collaboration whose work is discussed later in this chapter) remarked that there were, in fact, circumstances in which nonblind analysis would be preferred (Schwartz, 1995):

While a blind analysis does yield unbiased upper limits, it has one serious drawback: it is possible to miss an obvious background, subsequently observe a large number of candidates, and end up setting a very weak upper limit. This situation does a disservice to the experiment as the full "discriminating power" of the detector has not been used. Thus, one ultimately *should* look at events in the signal regions—after all cuts have been fixed—to check whether they are due to some trivial background or instrumental problem such as the high voltage having been tripped off. If such events can be attributed to such sources, then it makes more sense to cut them and set a biased but meaningful limit rather than leave them and set an unbiased but not useful limit. . . . Whichever choice is made, it is important that when publishing results one states exactly what was done and in what order the cuts were made, so that the reader can judge for him- or herself the significance of the limit set or the discovery made. (p. 2)

Some Examples

KTeV

The hidden offset method is similar to the technique suggested by Alvarez for the Fairbank experiment.[9] Its first use in recent high-energy physics was in experiments searching for rare decay modes of the K meson and for the weak decay of a hypothesized dibaryon (Arisaka et al., 1993a,b; Adler et al., 1996; Belz et al., 1996).

The method was also similar to that used by the KTeV collaboration in their search for direct CP violation in neutral K-meson decays (Alavi-Harati et al., 1999).[10] The parameter of interest was:

$$Re(\varepsilon'/\varepsilon) = 1/6[(\Gamma(K_L \to \pi^+\pi^-)/\Gamma(K_S \to \pi^+\pi^-))/(\Gamma(K_L \to \pi^0\pi^0)/ \Gamma(K_S \to \pi^0\pi^0))],$$

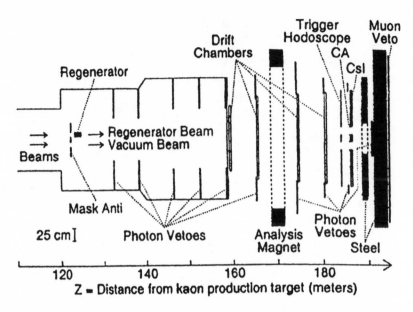

Figure 6.2. Plan view of the KTeV experimental apparatus configured to measure Re(ε'/ε). The label "CsI" indicates the electromagnetic calorimeter. From Alavi-Harati et al. (1999).

where the Γ's are the respective decay rates, the K_L and K_S are the long- and short-lived neutral K mesons, and the π's are pi mesons, or pions.

In any experiment, there will be many selection criteria and corrections to the data. Their final values will be set before the final analysis is done and they are thus independent of the ultimate result. Some criteria will initially be set quite loosely so that valuable data will not be lost. They may later be tightened as more detailed knowledge of the apparatus and the data is acquired. In addition, the final data sample may be quite sensitive to some of these criteria and not to others. Thus for example, the experimenters required that the reconstructed K-meson mass in the events be within ±10 MeV/c^2 of the accepted K-meson mass. No events would be lost due to this cut. In contrast, the requirement that no photon be within 7.5 cm of another photon in the $\pi^0\pi^0$ sample was guaranteed to eliminate good events, and thus the final result might very well be sensitive to this cut. The effect of such cuts was investigated in detail.

Let us examine the KTeV experiment to see some of these criteria. The experimental apparatus is shown in Figure 6.2. The experiment used two kaon beams from a single target. This allowed simultaneous data collection for both K_S and K_L decays. It also lessened the sensitivity of the experiment to time variations in the beam and in the detector efficiencies. The K_S beam was produced by placing a regenerator in the K_L beam.[11] The regenerator was on alternate sides of the apparatus for each accelerator beam extraction to minimize the effect of any left-right beam or detector asymmetry.

The quantity of interest $Re(\varepsilon'/\varepsilon)$ depends on the double ratio of $\pi^+\pi^-$ decays to that of $\pi^0\pi^0$ decays for K_S and K_L mesons. Thus it was crucial to know the relative acceptances for the different decay modes (Alavi-Harati et al., 1999):

To measure the double ratio of decay rates in the expression for $Re(\varepsilon'/\varepsilon)$, we must understand the *difference* between the acceptances for K_S versus K_L decays to each $\pi\pi$ final state. Triggering, reconstruction, and event selection are done with *identical criteria* [emphasis added] for decays in either beam, so the only major difference is in the decay vertex distributions, shown in Fig [6.3] as a function of Z, the distance from the kaon production target. Therefore the most crucial requirement for measuring $Re(\varepsilon'/\varepsilon)$ with this technique is a precise understanding of the Z dependence of the detector acceptance. (p. 23)

Because of the presence of the regenerator, the beginning of the allowed decay region was slightly different for the K_S and K_L beams:

In the regenerator beam [K_S] the decay region was defined by a lead-scintillator module at the downstream end of the regenerator. In the vacuum beam [K_L] the acceptance for decays upstream of $Z = 122$ m is limited by the "mask anti" (MA), a lead-scintillator counter with two square holes 50% larger than the beams. (p. 23; see Figure 6.2)

A detailed Monte Carlo simulation was used to calculate the detector acceptance for the $\pi\pi$ signal modes and to evaluate backgrounds.

The spectrometer included four drift chambers and a dipole magnet to measure the momentum of the charged particles produced in the K-meson decays, along with an electromagnetic calorimeter to measure the position and energy of the photons produced in the π^0 decay of the K mesons and the energy of charged particles. There was also a trigger hodoscope of scintillation counters. In addition, there were photon veto counters, a mask anticounter (MA) in front of the regenerator, and a col-

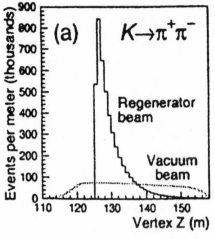

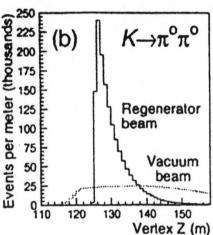

Figure 6.3. Decay vertex distributions for the (a) $K \to \pi^+\pi^-$ and (b) $K \to \pi^0\pi^0$ decay modes, showing the difference between the "regenerator" (K_S) and "vacuum" (K_L) beams. From Alavi-Harati et al. (1999).

lar anticounter surrounding each beam hole. These were designed to reduce background in the $\pi^0\pi^0$ channel, particularly from $K \to 3\pi^0$ decays. The regenerator itself was made of scintillator and viewed by phototubes, and served as a veto counter for interactions produced in the regenerator. Events were triggered by either synchronous signals in the trigger hodoscope ($\pi^+\pi^-$ decays) or by a fast-analog energy sum from the electromagnetic calorimeter ($\pi^0\pi^0$ decays).[12]

Cuts were also taken on the individual events. For example, for $\pi^+\pi^-$ events, each pion was required to have a momentum of at least 8 GeV/c and to deposit <85% of its energy in the calorimeter. The reconstructed $\pi^+\pi^-$ mass was required to be between 488 and 508 MeV/c^2. (The accepted mass of the neutral K meson is 497.672 MeV/c^2.) In addition, the square of the transverse momentum (p_T^2) of the $\pi^+\pi^-$ system relative to the initial kaon trajectory was required to be <250 MeV^2/c^2. These criteria were intended to ensure that the decay was due to a K meson in the beam.

For $\pi^0\pi^0$ candidates, four photons were detected by the calorimeter. The photon pairing combination that was most consistent with hypothesis of two π^0 decays at a common point (each π^0 meson decays into two photons in approximately 10^{-16} s). This was interpreted as the kaon decay vertex. Each photon was required to have an energy of at least 3 GeV, to be about 5 cm from the outer edge of one of the CsI counters, and to be 7.5 cm from any other photon. The reconstructed four-photon mass was required to be between 490 and 505 MeV/c^2. To guard against kaon scattering, the energy centroid of the four photons at the CsI counters that comprised the electromagnetic calorimeter was used to calculate a "ring number." This was defined as four times the square of the larger normal distance (horizontal or vertical) in centimeters from the energy centroid to the closest beam. The ring number was required to be <110, which selected events within a square of 110 cm^2 centered on each beam. Other cuts were made on energy deposits in MA, photon veto counters, and regenerator.

These selection criteria were determined and then applied to construct the final data sample from which the value of Re(ε'/ε) would be calculated. They were set before that quantity was calculated and were thus independent of its value.

Some of the data obtained are shown in Figure 6.4. The cuts on p_T^2 and on ring number are shown by arrows. The Vac $\pi\pi$ data show the K$_L$ decays, and the Reg $\pi\pi$ data show K$_S$ decays. The signal is the data minus the background:

Background contributions to the $\pi^+\pi^-$ samples are determined by using the sidebands in the mass and p_T^2 distributions to normalize MC [Monte Carlo] predictions for the various background processes. Fig. [6.4](a) and [6.4] (b) show that the p_T^2 distributions for data are well described by the sum of coherent $\pi\pi$ MC and total background MC. (Alavi-Harati et al., 1999, p. 24)

For the larger $\pi^0\pi^0$ background, a Monte Carlo calculation of the background was normalized by using the region where no signal was expected (ring number 286–792). (Note here the importance of Monte Carlo

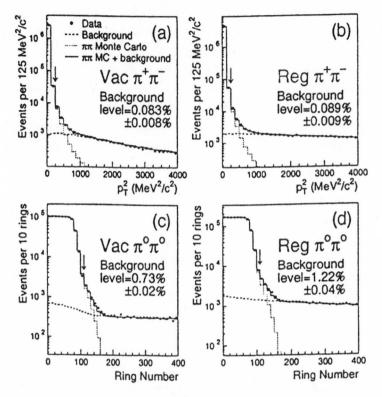

Figure 6.4.(a,b) Distributions of p_T^2 for the $\pi^+\pi^-$ samples. (c,d) Ring number for the $\pi^0\pi^0$ samples. Total uncertainties are given for the samples passing the analysis cuts (arrows). From Alavi-Harati et al. (1999).

calculations in producing the result). "After background subtraction, the net yields are 2 607 274 $\pi^+\pi^-$ in the vacuum beam, 4 515 928 $\pi^+\pi^-$ in the regenerator beam, 862 254 $\pi^0\pi^0$ in the vacuum beam and 1 433 923 $\pi^0\pi^0$ in the regenerator beam" (Alavi-Harati et al., 1999, p. 25). The experiment that first demonstrated the existence of the decay $K_L \rightarrow \pi^+\pi^-$, and thus, CP violation, had a signal of 45 ± 9 events (Christenson et al., 1964). Things had improved considerably.

The experimenters obtained their value of Re(ε'/ε) from analysis of the decay distributions of the K meson decays. "Re(ε'/ε) is extracted from the background subtracted data [Figure 6.5] using a fitting program [the program fitted 24 different parameters]. . . . *Fitting was done "blind," by hiding the value of Re(ε'/ε) with an unknown offset, until after the analy-*

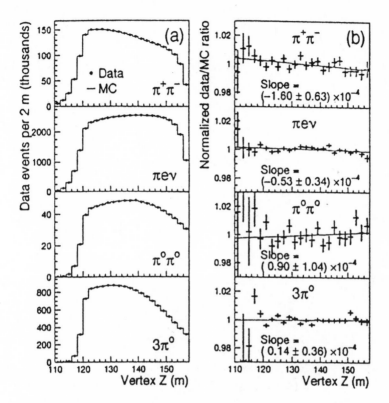

Figure 6.5. (a) Data versus Monte Carlo comparisons of vacuum-beam Z distributions for $\pi^+\pi^-$, $\pi e\nu$, $\pi^0\pi^0$, and $3\pi^0$ decays. (b) Linear fits to the Data/Monte Carlo ratio of Z distributions for each of the four samples. From Alavi-Harati et al. (1999).

sis and systematic error evaluation were finalized" (Alavi-Harati et al., 1999, p. 25; emphasis added). The analysis procedures and parameters were determined without the experimenters knowing the value of $\mathrm{Re}(\varepsilon'/\varepsilon)$. The group also worried about the possibility that the cuts applied in their analysis might have affected their value of $\mathrm{Re}(\varepsilon'/\varepsilon)$. "We assign systematic errors based on the dependence of the measured value of $\mathrm{Re}(\varepsilon'/\varepsilon)$ on variations of key analysis cuts, in particular the p_T^2 cut for the $\pi^+\pi^-$ and the ring-number and photon quality cuts for $\pi^0\pi^0$. No significant dependence on other analysis cuts is observed" (p. 25).

The KTeV group had also investigated the robustness of their result.

We have performed several cross-checks on the $\mathrm{Re}(\varepsilon'/\varepsilon)$ result. Consistent values are obtained at all kaon energies, and there is no significant variation as a

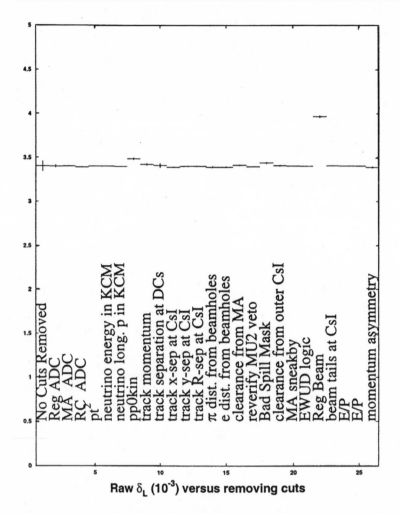

Figure 6.6. Raw value of the decay asymmetry ($\delta_L \times 10^{-3}$) versus removing cuts. From Nguyen (2001).

function of time or beam intensity. . . . We have also extracted $Re(\varepsilon'/\varepsilon)$ using an alternative fitting technique which compares the vacuum- and regenerator-beam Z distributions directly, eliminating the need for a Monte Carlo simulation to determine the acceptance.[13] While less statistically powerful, this technique yields a value of $Re(\varepsilon'/\varepsilon)$ which is consistent with the standard analysis."[14] (p. 27)

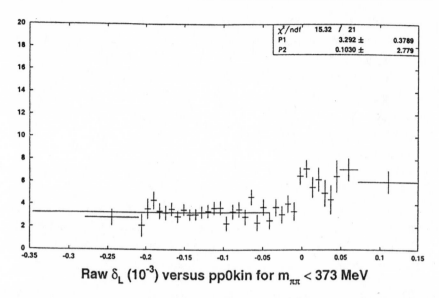

Figure 6.7. Raw value of the decay asymmetry ($\delta_L \times 10^{-3}$) versus pp0kin for $m_{\pi\pi}$ < 373 MeV. From Nguyen (2001).

Tests for robustness of a result were also illustrated in the KTeV measurement of δ_L, the charge asymmetry in K^0_{e3} decay. (There is a difference between the decay rates for $K^0_L \rightarrow e^-\pi^+\bar{\nu}_e$ and that of $K^0_L \rightarrow e^+\pi^-\nu_e$.) Figure 6.6 shows the raw value of δ_L (before corrections and unblinding) versus removing various cuts made on the data. One can see that with the exception of two of the cuts "pp0kin" and "Regbeam," the value of δ_L is constant, within statistics, for all of the cuts.

"Pp0kin" is a kinematic cut designed to remove events due to $K^0_L \rightarrow \pi^-\pi^+\pi^0$ decays, which form a significant background to the electron decay process being studied.[15]

The $\pi^-\pi^+\pi^0$ [charged] pions would have to be misidentified as electrons (E/P > 0.925) and this is a very charge asymmetric process.[16] This can be seen in Figure [6.7] where the result in the $\pi^-\pi^+\pi^0$ background region is systematically higher than in the background free region. (Nguyen, 2001, p. 50)

The experimenters wanted to accept only those events that originated in the vacuum beam. (Recall that the experiment included two beams, a vacuum beam and a regenerator beam; see Figure 6.2). The cut on the regenerator beam reduced events produced in the beam that passed through the two-meter-long regenerator. Passing through the regenerator scintillating material produces background events with a very different asymmetry than that of the K_{e3}^0 decay being studied. With the regenerator beam cut in place, the number of events due to crossover events (those from the wrong beam) was negligible.

The final value found for δ_L (the asymmetry in K_{e3}^0 decay) was (3320 $\pm$ 74) $\times$ 10^{-6}. The accepted value given by the Particle Data Group was (3330 $\pm$ 140) $\times$ 10^{-6} (Groom et al., 2000, p. 524). The agreement is remarkably good. In private conversation, several of the experimenters remarked that such good agreement might have been suspect had blind analysis not been performed. The value of δ_L was not known until after all of the cuts and corrections had been finalized.

E791: Comparing Blind and Nonblind Analysis

An interesting question is whether a blind analysis provides a result different from that of an ordinary analysis when both procedures are applied to the same data. There is an episode from the history of recent physics in which the same data were analyzed using both types of analysis:[17] the search for rare decay modes of the D meson by the Fermilab E791 collaboration. The decays studied in the first analysis were $D^+ \rightarrow \pi^+\mu^+\mu^-$ and $D^+ \rightarrow \pi^+e^+e^-$ (Aitala et al., 1996), whereas the second, more extensive, analysis included both those decays and 22 others (Aitala et al., 1999). It is clear that the experimenters were concerned about the possibility of experimenter bias. The selection criteria for all D^+ decays were determined without examining the decays of interest. "To search for the D^+ FCNC [flavor-changing, neutral-current] decays in an *unbiased* way, the track and vertex selection criteria for all D^+ decays were determined from the Cabibbo-suppressed mode $D^+ \rightarrow \pi^-\pi^+\pi^+$" (Aitala et al., 1996, p. 365, emphasis added). Other criteria, event selection, track matching, and minimum momentum were "selected in an *unbiased* way by optimizing $S_{MC}/\sqrt{B_{data}}$, where S_{MC} is a Monte Carlo FCNC signal and B_{data} is the background of misidentified hadrons in data *outside the FCNC signal region*" (p. 365, emphasis added).

The experimental apparatus is shown in Figure 6.8. The search for $D^+ \rightarrow \pi^+\mu^+\mu^-$ and $D^+ \rightarrow \pi^+e^+e^-$ decays also required muon and electron identification criteria, respectively. Such criteria were set independently of the final result. The muons were identified by scintillation

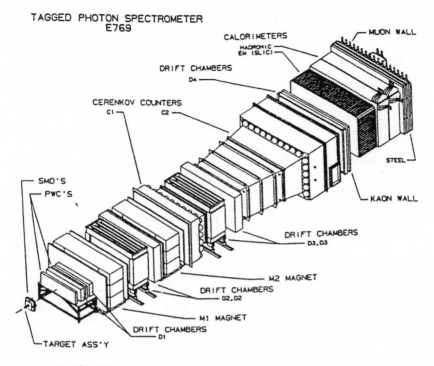

TAGGED PHOTON SPECTROMETER
E769

Figure 6.8. The E791 experimental apparatus. From Appel (1992).

counters located behind 15 interaction lengths of shielding. Muons have a longer range in matter than either pions or electrons, and the probability of one of those particles penetrating the shielding was very low. The muon counter efficiency was measured in special runs using independent muon identification and was found to be 99 ± 1%. Electrons were identified by the lead and liquid scintillator calorimeter. The identification was based on energy deposition, shower shape, and position in the calorimeter. "Calorimeter response was studied with topologically identified electron-positron pairs from conversions upstream of the tracking, and with pions from kinematically identified $K_S^0 \rightarrow \pi^+\pi^-$ decays" (Aitala et al., 1996, p. 366).

The results of the search for $D^+ \rightarrow \pi^+\mu^+\mu^-$ are shown in Figure 6.9. The upper curve was obtained with no identification of the decay particles as muons. It was fit quite well by a combination of Gaussian peaks due to misidentified $D^+ \rightarrow \pi^-\pi^+\pi^+$ and $D_S^+ \rightarrow \pi^-\pi^+\pi^+$ added to an exponential background.[18] The lower curve includes muon identifica-

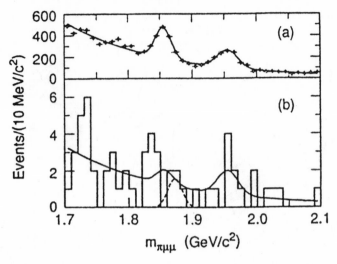

Figure 6.9. Search for a $D^+ \to \pi^+\mu^+\mu^-$ signal. (a) Invariant-mass spectrum assuming a $\pi^+\mu^+\mu^-$ hypothesis but with no muon identification requirement (diamonds). The curve, which is a fit by the sum of Gaussian peaks from misidentified $D^+ \to \pi^-\pi^+\pi^+$ and $D_S^+ \to \pi^-\pi^+\pi^+$ and an exponential background, determines the central values and widths of the peaks. (b) The $\pi^+\mu^+\mu^-$ invariant-mass spectrum for events with muon identification (histogram). The solid curve is a best fit to contributions from $D^+ \to \pi^+\mu^+\mu^-$, feedthrough from $D^+ \to \pi^-\pi^+\pi^+$ and $D_S^+ \to \pi^-\pi^+\pi^+$, and an exponential background. The dashed curve shows the size and shape of the $D^+ \to \pi^+\mu^+\mu^-$ contribution ruled out at the 90% confidence level. From Aitala et al. (1996).

tion, which are the candidates for the decay mode, and were fit in the same way. The dashed curve shows the size and shape of the $D^+ \to \pi^+\mu^+\mu^-$ contribution ruled out at the 90% confidence level.

The result for the electron decay $D^+ \to \pi^+e^+e^-$ is shown in Figure 6.10. After candidate events (which included electron identification) had been found, a search window in the mass region 1.830–1.890 GeV/c^2 was chosen using a simulated decay signal.[19] Only one of three candidates for the decay was inside this window. The experimenters used the two candidate events outside the window to estimate the background inside the window and found a background of 0.42 ± 0.29 events, giving a signal of 0.58 ± 1.04 events. Application of a standard Poisson method resulted in 3.56 events as the upper limit for the decay.

The experimenters concluded, "In summary, Fermilab experiment E791 has obtained upper limits on branching fractions B for the

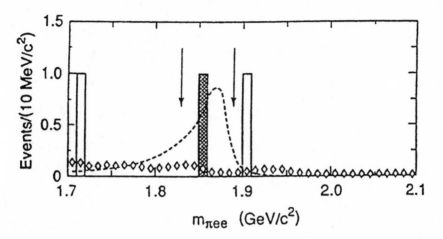

Figure 6.10. Search for $D^+ \to \pi^+e^+e^-$ signal. Invariant-mass spectrum with $\pi^+e^+e^-$ hypothesis. Three events pass the electron-identification requirements (histogram). One of them is in the signal region between the arrows. Background is estimated from the $\pi^+e^+e^-$ invariant-mass spectrum without the electron-identification requirement, normalized to two events outside the signal region (diamonds). The dashed curve shows the size and shape of the bremsstrahlung-widened $D^+ \to \pi^+e^+e^-$ signal excluded at the 90% confidence level. From Aitala et al. (1996).

three-body FCNC decays $D^+ \to \pi^+\mu^+\mu^-$ and $D^+ \to \pi^+e^+e^-$ that are an order of magnitude below those previously published. At 90% C.L. [confidence level], $B(D^+ \to \pi^+\mu^+\mu^-) < 1.8 \times 10^{-5}$ and $B(D^+ \to \pi^+e^+e^-) < 6.6 \times 10^{-5}$" (Aitala et al., 1996, p. 367). These results were accepted by the Particle Data Group as the definitive limits.[20]

The group also varied their selection criteria and showed that their result was robust against reasonable variations in these criteria. For example, in studying the decay $D^+ \to \pi^+\mu^+\mu^-$, the experimenters observed (Aitala et al., 1996):

Variations on this technique gave consistent results. Specifying widths of the expected FCNC signal between 11 and 15 MeV/c^2 changes the upper limit by only 4%. Constraining the relative amounts of $D^+ \to \pi^-\pi^+\pi^+$ and $D_s^+ \to \pi^-\pi^+\pi^+$ feedthrough to be the same as in Fig. [6.9] gives < 3.2 events at the 90% C.L., while use of a simple mass window instead of a likelihood fit gives < 4.5 events. . . . At 90% C.L., $B(D^+ \to \pi^+\mu^+\mu^-) < 1.8 \times 10^{-5}$. The limit is quite stable under variation of vertex selection and muon ID criteria. (p. 366)

Another interesting question that arises in an experiment that obtains a null result is whether the experiment would actually have detected the phenomenon in question had it been present.[21] The experimenters checked this by inserting simulated events into their observed distribution. "We have also tested the procedure with ensembles of simulated experiments in which fixed numbers (2–10) of simulated FCNC signal events, drawn randomly from a Gaussian mass distribution, are added to the observed spectrum and *successfully found by the fit*" (p. 366, emphasis added). If the decays had been present they would have been detected.

The second paper on rare decays published by the E791 collaboration "blindly" analyzed the same data and appeared in 1999 (Aitala et al., 1999):

For this study we used a "blind" analysis technique. Before our selection criteria were finalized, all events having masses within a window ΔM_S around the mass of the D^+, D_S^+, or D^0 were "masked"[22] so that the presence or absence of any potential signal candidates would not bias our choice of selection criteria. All criteria were then chosen using signal events generated by a Monte Carlo simulation program and background events from real data. Events within the signal windows were unmasked only after this optimization. Background events were chosen from a mass window ΔM_B above and below the signal window ΔM_S. The criteria were chosen to maximize the ratio $N_S/\sqrt{N_B}$ where N_S and N_B are the numbers of signal [generated by the Monte Carlo simulation] and background events, respectively. (pp. 403–4)

Although the data used remained the same, other criteria were changed slightly as detailed study of the apparatus and the data continued in the intervening three years. In addition, somewhat different statistical methods, unavailable earlier, were used to calculate the upper limits. The final upper limits for the decay modes $D^+ \rightarrow \pi^+\mu^+\mu^-$ and $D^+ \rightarrow \pi^+e^+e^-$ were $<1.5 \times 10^{-5}$ and $<5.2 \times 10^{-5}$, respectively. These were not substantially different from the limits of $<1.8 \times 10^{-5}$ and $<6.6 \times 10^{-5}$ reported earlier and no significance was attached to the difference.

One point of interest is that the number of events in the final sample of $D^+ \rightarrow \pi^+e^+e^-$ candidates had changed. Figure 6.10 shows three events, one in the signal region and two background events. In the 1999 publication, the higher-mass background event is missing (Figure 6.11). Conversations with several of the experimenters revealed that the missing background event had not, in fact, been noticed. They attributed its loss to the

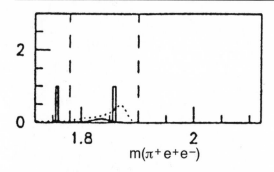

Figure 6.11. Final sample of candidate events for $D^+ \rightarrow \pi^+ e^+ e^-$. The solid curves are the estimated background; the dotted curve represents the signal shape for the number of events equal to the 90% confidence level upper limit. The dashed vertical lines are the M_S boundaries. From Aitala et al. (1999).

slightly different selection criteria used in the two papers and noted that the upper limit for the decay had not changed significantly.[23]

The Anomalous Magnetic Moment of the Muon

Blind analysis has become so widespread that there are now instances in which several subgroups of an experimental group perform independent blind analyses of the same experimental data. This provides a strong safeguard against experimenter bias. An example of this is the recent precision measurement of the anomalous magnetic moment of the muon (Brown et al., 2001). This is a quantity that is precisely calculable by the Standard Model, the currently accepted theory of elementary particles, and any discrepancy between the experimental and theoretical values would be evidence that the current theory needs modification or replacement.

In this experiment, positive polarized muons (their spins are aligned) were injected into a storage ring, in which the muons were kept in orbit by a magnetic field $<B>$. The muon spin precesses faster than its momentum rotates in the magnetic field by an angular frequency ω_a. The anomalous magnetic moment a is:

$$a_\mu = \omega_a / [(e/m_\mu c) < B >].$$

The angular frequency ω_a was determined by counting the number of positrons resulting from muon decay, $\mu^+ \rightarrow e^+ + \nu_e + \bar{\nu}_\mu$. Parity vio-

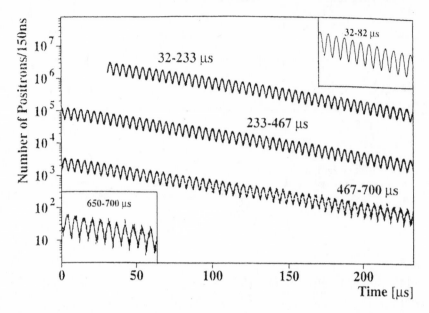

Figure 6.12. Positron time spectrum overlaid with the fitted ten-parameter function (χ^2/DOF = 3818/37990). The total event sample of 0.95 × 10^9 e$^+$ with E ≥ 2.0 GeV is shown. From Brown et al. (2001).

lation in the decay[24] gives rise to asymmetries in both the electron angular distribution and the electron energy. The number of positrons with energy >E is:

$$N(t) = N_0 e^{-t/(\gamma\tau)} \{1 + A(E) \sin[\omega_a t + \varphi_a(E)]\},$$

where $\gamma\tau$ is the time dilated lifetime of the muon, A(E) is a function of the positron energy, and $\varphi_a(E)$ is a phase depending on the positron energy. Thus by measuring the distribution of decay positrons, one can determine ω_a (Figure 6.12). The other crucial parameter in determining a_μ is the magnetic field <B>. The magnetic field was measured by 17 nuclear magnetic resonance (NMR) probes mounted on a trolley that moved on a fixed track inside the muon storage ring vacuum chamber. The measurements were made approximately every three days, and interpolation for the period between measurements was provided by 150 fixed NMR probes distributed around the ring. Because of its importance, the fitting of the magnetic field was delegated to four different and independent subgroups.[25] Each group used a different random offset in the value of the magnetic field. One of the groups had difficulty in obtaining an internally consistent fit to the field; in an effort to locate the problem, the analysis was partially unblinded. Each group was then given the same random offset, allowing a comparison of the fits. The problem was found and corrected. The absolute value of the field, needed for the calculation of a_μ, was still unknown to the experimenters.

Interestingly, the experiment-theory comparison was, in a sense, also blind. After the experimental group had obtained their final value of a_μ, they asked William Marciano, a theoretical physicist who worked on calculating its value, what the best theoretical value of a_μ was. They did not inform him of their result until after he had provided the value. Their experimental value was $a_\mu = 11{,}659{,}202\ (14)(6) \times 10^{-10}$, which was in good agreement with previously measured values, but it has an experimental uncertainty only one-third the size.[26] The best theoretical value, calculated from the Standard Model, was $a_\mu = 11{,}659{,}159.6(6.7) \times 10^{-10}$. The difference $a_\mu\ (\text{exp}) - a_\mu\ (\text{SM}) = 43(16) \times 10^{-10}$ might indicate a problem with the Standard Model, and is, perhaps, an indication of the presence of supersymmetry, a currently favored theoretical speculation.

Discussion

It is interesting to consider whether blind analysis could have been applied to the cases of selectivity we have already discussed: the measurement of the K_{e2}^+ branching ratio, Millikan's measurement of the charge of the electron, the search for gravity waves, the 17-keV neutrino, and the search for low-mass electron-positron states.

In the case of the K_{e2}^+ branching ratio, the analysis was effectively blind. The cuts on range, tracking matching, and time of decay were all fixed and their robustness checked before the number of K_{e2}^+ candidates was normalized to other, more prevalent decay modes.[27] For Millikan's experiment, one suspects that a random offset added to his calculated value of e for each event might have guarded against his selectivity on the oil drops used. It would not, however, have eliminated his selective calculational methods. Assuming the same random offset for each oil drop, the two methods would still have given different values of e. As we have seen, however, the correctness of Millikan's result was shown by the numerous repetitions of the experiment.

For the 17-keV neutrino, it might have been possible to add the same random offset to each measured electron energy. In that way, the energy at which the kink in the energy spectrum occurred would not have appeared at the energy that gave a 17-keV neutrino, but at some other energy. This might have avoided any possible bias toward reproducing previous results.[28] In the case of Weber's gravity waves, it is not clear to me how one might have applied blind analysis. In the episode of the low-mass electron-positron states, one might argue that had the same random offset been applied to the electron and positron energies, any bias toward reproducing the previously observed lines in the sum-energy spectrum would have been eliminated. The "hidden signal box" method could not have been used because the entire sum-energy spectrum was regarded as the signal region. There was no signal-free region in which the value of the cuts could have been fixed and then applied to the signal region. There was also the very real possibility that the observed effects were extremely sensitive to the experimental conditions. As discussed earlier in the book, the artifactual nature of the results in the last three episodes was shown by normal strategies. Blind analysis was not needed.

Judging from the significant amount of effort being devoted to blind analysis, it seems clear that the problem of selectivity or experimenter bias is one that troubles the physics community. Blind analysis is a proactive strategy designed to eliminate or minimize the problem. It is not, as we have seen, applicable to all experiments. It is yet another strategy used to argue for the correctness of experimental results. As we have discussed, the robustness of the results was checked in the standard way in the experiments that used blind analysis.

Monte Carlo Simulation

In several of the episodes discussed earlier, including the early searches for gravity waves, the 17-keV neutrino, the low-mass electron-positron states, and several of the experiments discussed in the section on blind analysis, Monte Carlo calculations or simulations played an important role.[29]

Andrew Pickering has questioned the use of such Monte Carlo calculations and suggested that their use in experiments precludes the use of the results as evidential support. In discussing the use of such a simulation in the Gargamelle experiment, which reported the existence of weak neutral currents, Pickering noted that several of the inputs to the calculation could be questioned. These included the beam characteristics, the interaction of nucleons with atomic nuclei, neutron production, and idealized experimental geometry (Pickering, 1984b):

My object here is simply to demonstrate that assumptions were made which could be legitimately questioned: one can easily imagine a determined critic taking issue with some or all of these assumptions. Moreover, even if all of the assumptions were granted, it remained the case that they were input not to an analytic calculation, but to an extremely complex numerical simulation. The details of such simulations are enshrined in machine code and are therefore inherently unpublishable and not independently verifiable. Thus the skeptic could legitimately accept the input to the calculation but continue to doubt its output. (p. 96)

What Pickering overlooks is that considerable effort is devoted to checking the results of that calculation by comparison with experimental evidence that is independent of the result in question.[30] The results of this checking are, in fact, publicly available in the published literature. Thus in the 17-keV neutrino episode, Hime's (1993) Monte Carlo calculation had shown that intermediate scattering effects in his aluminum baffles could account for his data just as well as did the assumption of a 17-keV neutrino. He checked his calculation by comparing it to data taken with the same experimental apparatus and geometry using a monoenergetic internal conversion electron source. The excellent fit between these measurements and his simulation argued for the correctness of his calculation (Figure 6.13).

Such checks of simulations are usually done. For example, in an experiment designed to measure the energy dependence of the form factor in K_{e3}^{+} decays, $K^{+} \rightarrow e^{+} + \pi^{0} + \nu$, the way in which the energy-dependent parameter λ was fixed was by comparing Monte Carlo-generated spectra that used different values of λ with experimental data (Imlay et al., 1967). The Monte Carlo simulation was checked by comparing its results with a sample of background events:

It was also necessary to know the energy distributions relating to background events. These distributions were obtained from the Monte-Carlo generated sample of spurious K_{e3}^{+} events. Indications of the validity of this calculation were obtained from the distributions of positron momentum, γ-ray energy, and π^{0}

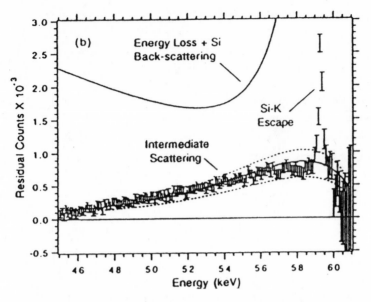

Figure 6.13. ^{109}Cd spectrum accumulated in the Oxford geometry. The solid curve shows the effect calculated for intermediate scattering. From Hime (1993).

energy for those *events which were rejected by selection criterion 3.* This criterion required that the counter behind each spark chamber give a pulse if the shower in the chamber contained sparks in either of its last two gaps. These rejected events should differ from the background events in the final sample of 1867 nominal K^+_{e3} events only with regard to selection criterion 3. Thus, when reconstructed as K^+_{e3} decays, the background events that passed and failed criterion 3 should have exactly the same distributions. These are shown in Fig. [6.14], along with the calculated distributions for Monte-Carlo generated spurious events. The good agreement provides strong support for the background calculation, particularly since these distributions differ substantially from the corresponding distributions for good events. (p. 1209)

Peter Galison's (1987, Chapter 4) discussion of the Fermilab E1A experiment, which argued for the existence of weak neutral currents, also illustrates the checking of a Monte Carlo calculation. In this experiment, charged-current events were those that included a muon (which was identified by its ability to penetrate an absorber) as well as a hadron (i.e., strongly interacting particle) shower. Neutral-current events contained

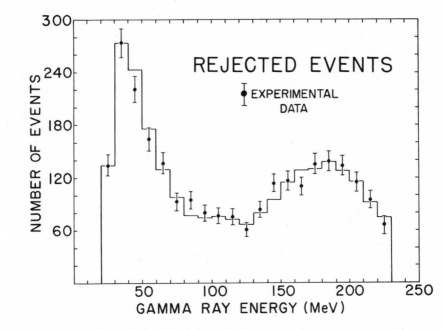

Figure 6.14. Comparison of the Monte Carlo-generated γ-ray spectrum with experimental data for rejected events. From Imlay et al. (1967).

the hadron shower, but did not include a muon. One crucial problem was that hadrons in a neutral-current event might penetrate (punch through) the absorber and be classified as charged-current event. Thus, the estimate of hadron punchthrough was crucial in establishing the existence of the neutral current. The experimenters used a Monte Carlo calculation to estimate the punchthrough. They checked their simulation by rescanning 30% of their data and measuring the fraction of charged current events that also contained a hadron punchthrough. They plotted the punchthrough probability as a function of energy and found that it matched the results given by the Monte Carlo calculation. This increased their confidence in their model and they applied the computer simulation to calculate the number of neutral current events. The experimenters also analyzed their data using a different method and obtained the same result, showing the robustness of both the result and the correctness of the Monte Carlo calculation. "[W]ithin a week or so [we] will have the measurements needed to analyze the data using SC4 alone in a nearly punch-through-free way. Since we also measure [the punchthrough probability] . . . with almost no reliance on Monte Carlo, we would like to wait to include this analysis in the paper" (Galison, 1987, p. 237). In addition, the Monte Carlo calculation was done independently by two different subgroups within the experimental group; the results were indistinguishable.

Pickering also overlooks the fact that the robustness of the results of a Monte Carlo calculation is checked against reasonable variations in the simulation input parameters. This is done because—as Pickering himself notes—these parameters are not known exactly. Typically, the results are not sensitive to such variations. If they are, then the results must be used with extreme care, and may not, in fact, be usable.

Such independent checks of the robustness of Monte Carlo simulations instill confidence in the calculations in all but the most determined skeptics.

II THE RESOLUTION OF DISCORDANT RESULTS

As we have seen in Part I, experimental results often disagree. How, then, can scientific knowledge be based on experiment? If we do not have good reasons for belief in experimental results or for our choice of one of a set of discordant results rather than another, then experimental evidence cannot provide the grounds for scientific knowledge. Although in practice, the discord between experimental results is usually resolved within a reasonable time,[1] questions remain as to whether the method by which the resolution is achieved provides grounds for confidence in the knowledge gained from the experimental results.

Social constructivists, whose views we discussed in the Introduction, imply (however much they may disclaim it)[2] that it does not. In their view, the resolution of such disputes and the acceptance of experimental results in general is based on "negotiation" within the scientific community, which does not include epistemological or methodological criteria. Such negotiations do include considerations such as career interests, professional commitments, prestige of the scientists' institutions, and the perceived utility for future research. As Pickering (1984b) stated, "Quite simply, particle physicists accepted the existence of the neutral current because they could see how to ply their trade more profitably in a world in which the neutral current was real" (p. 87). The emphasis on career interests and future utility is clear.[3]

Collins (1985) has summed up the argument against both experimental results and reasoned resolution of discordant results in what he calls the "experimenters' regress": What scientists take to be a correct result is one obtained with a good, that is, properly functioning, experimental apparatus. But a good experimental apparatus is simply one that gives correct results. Collins claims that there are no formal criteria that one can apply to decide whether an experimental apparatus is working properly. "Proper operation of the apparatus, parts of the apparatus and the experimenter are defined by the ability to take part in producing the proper experimental outcome. Other indicators cannot be found" (Collins, 1985, p. 74).

I disagree. I believe that the discord between experimental results is resolved by reasoned argument, based on epistemological and methodological considerations. These are the other indicators. This does not preclude a joint decision concerning whether a detector works properly and the phenomenon in question exists. The disagreement between my view and that of the constructivists concerns the reasons for that decision. As we have seen and as discussed below, the decision between the discordant results in the episodes of gravity waves, the 17-keV neutrino, and the proposed low-mass electron-positron states was based on epistemological and methodological criteria.

Some commentators (as well as social constructivists themselves) have argued that constructivists do not claim that scientists do not provide reasons for their decisions, but rather that the reasons are insufficient. "Social constructivists do *not* say that experimental evidence is irrelevant to theory choice, confirmation, or refutation. Nor do they argue that there are no good reasons for belief in the validity of evidence" (Lynch, 1991, pp. 476–77). Nevertheless, in studies presented by constructivists, evidence does not enter into such decisions, nor are good reasons for belief in evidence discussed. The contructivist claim is twofold. First, the reasons used by the scientific community do not provide justification for either experimental evidence or for hypothesis testing on the basis of that evidence. Second, even if such reasons were sufficient within science, they do not have any standing beyond the scientific community.

I shall begin with the second point. As discussed earlier, I believe that there is an epistemology of experiment, a set of strategies that provides grounds for reasonable belief in experimental results. I have further argued that these strategies have independent philosophical justification and have shown that they are used by scientists. Decisions between discordant results are made by the community of scientists, and are thus inherently social and dependent on historical context, particularly on what is accepted as scientific knowledge at a given time. I certainly do not deny that scientists have the usual human motivations, such as career advancement, desire for credit, prestige, and economic gain. Scientists also have an interest in producing scientific knowledge, as well as a career interest in producing correct results. I do claim, however, that such decisions are based on epistemological and methodological criteria, and that these criteria are not justified merely by their acceptance by the scientific community.

I believe that we have independent grounds for believing that science and its methodology provide us with reliable knowledge about the world. It is not just the successful practice of science, which is, after all, decided by scientists themselves, but rather evidence from the "real" world that underlies this judgment. It is not mystical incantations by Faraday, Maxwell, or other scientists that cause a light to come on when a switch is thrown. The earth would not suddenly head toward outer space if the American Physical Society voted to repeal Newton's law of universal gravitation. These and other examples too numerous to mention provide grounds for believing that science is actually telling us something reliable about the world. As Ian Hacking said (in the more limited context of discussing the reality of scientific entities such as electrons), "We are completely convinced of the reality of electrons when we regularly set out to build—and often enough succeed in building—new kinds of devices that use various well-understood causal properties of electrons to interfere in other more hypothetical parts of nature" (Hacking, 1983, p. 265). It is this practical intervention in the world that persuades us that we should take account of what 20th-century physics has to say when we formulate a world view. It is possible that negotiations based on the considerations suggested by some constructivists might give us reliable knowledge about the world, but that seems rather unlikely. Why should the world be such that it benefits the career interests of scientists?

The first point made by constructivists, concerning whether reasons are sufficient to provide justification for evidence or theories, relies on two philosophical points (Nelson, 1994). The first is the underdetermination of theory by evidence—the fact that one can always find an alternative explanation for a given experimental result. The second is the Duhem-Quine thesis: If an experiment seems to refute a theory, it in fact refutes the conjunction of both the theory and background knowledge and one does not know where to place the blame for the failure. One may save a hypothesis from refutation by suitable changes in one's background knowledge. I believe that adequate answers have already been provided for these points (e.g., Franklin, 1990, pp. 144–61; Franklin, 1993b, pp. 260–67). An adequate discussion of these issues would take us too far from the central issue of this section, namely: How is the discord between experimental results resolved?

I have argued for the existence of an epistemology of experiment, a set of strategies that can be used to defend the correctness of an experimental result. The difficulty is that in cases of discordant results, such strategies were applied to each of the experiments. The resolution must proceed by demonstrating that, in at least some of the experiments, the strategies have been incorrectly applied.

Perhaps the most important method of invalidating a result is to show that the Sherlock Holmes strategy has been incorrectly applied. One can argue that the experimental result can be explained by an alternative hypothesis, or that a plausible source of error (e.g., a background that might either mask or mimic the correct result) has been overlooked. One can demonstrate that the use of a particular strategy generates a contradiction with accepted results. Similarly, one might examine the assumptions concerning the operation of the apparatus and show empirically that they are incorrect. Plausible interpretations of the results may also be shown to be incorrect.

Other criteria can also be used. In a particular experiment, some epistemological strategies may have been applied successfully, whereas others had failed, casting doubt on the result. Sometimes the failure to reproduce an observation, despite numerous attempts to do so, might be legitimately regarded as casting doubt on the original observation, even when no error has been found in the original experiment. This would be a case of preponderance of evidence.

There are several different types of discordant experimental results. One may have experiments that measure the same quantity with the same, or similar, types of apparatus (e.g., the early searches for gravity waves, discussed in Chapter 2). Discordant results may also involve the measurement of the same quantity, but with different types of experimental apparatus (e.g., the 17-keV neutrino, discussed in Chapter 4). In this case, one might worry that the difference in the results is due to some crucial difference in the apparatus. A third type of discord occurs when different experiments, measuring different quantities that are predicted by the same theory, give results such that one of the experiments confirms the theory, whereas the other confutes theoretical prediction. One may even have discordant results produced when different members of the same experimental group analyze the same data in different ways. In addition, I present an episode in which two experimental results were regarded as discordant, but were, in fact, both correct. The discord resulted from an error in the interpretation of one of the results. Each of these types of discord is illustrated in chapters in Part II.

In these chapters, I examine four episodes from the recent history of physics: the suggestion of a Fifth Force, a modification of Newton's law of gravitation; early experiments on the absorption of β particles; experiments on neutrino oscillations; and experiments on atomic parity violation and the scattering of polarized electrons, and their relation to the Weinberg-Salam unified theory of electroweak interactions. In each of these episodes, discordant results were reported, and a consensus was subsequently reached that one result, or set of results, was incorrect. I examine the process of reaching that consensus, and that this process was based on the epistemological and methodological criteria I have suggested.

Can case studies be used to demonstrate that scientists resolve the discord between experimental results by the application of epistemological and methodological criteria? The case studies show only that in the six episodes presented here, the discords indeed were so resolved. Nevertheless, case studies do support the generalization. Although it is dangerous to generalize from only seven instances, I believe that these episodes provide a reasonable picture of the practice of modern physics. But note that constructivists also provide case studies to support their view of science. Two of the episodes discussed in this book—namely, the early attempts to detect gravity waves and the atomic-parity violation experiments—have been used by constructivists to support their view that the resolution of such discordant results does cast doubt on the status of science as knowledge (Collins, 1985; Pickering, 1984a, 1991). I have argued in detail elsewhere that their accounts are incorrect (Franklin, 1990, 1993c, 1994). Constructivists such as Pickering and Collins seem to imply that epistemological criteria are never decisive in resolving the dispute between discordant results. In that case, the presentation of even one case study in which the criteria are decisive casts doubt on their view.

7

The Fifth Force

In January 1986, Aronson, Fischbach, and Talmadge proposed a modification of Newton's law of universal gravitation (Fischbach et al., 1986). The Newtonian gravitational potential is $V = -Gm_1m_2/r$. Their modification took the form $V = -Gm_1m_2/r\ [1 + \alpha e^{-r/\lambda}]$, where α is the strength of the new interaction and λ is its range. This new interaction became known as the "Fifth Force." Their initial suggestion was that α was approximately 1% and λ approximately 100 m. Unlike the gravitational force itself, the new force was composition dependent. The Fifth Force between a copper mass and a platinum mass would be different from that between a copper mass and an iron mass. By early 1990, a consensus was reached that such a force did not exist. The decision process was not simple. There were two different sets of discordant results: (1) from measurements of gravity using towers and mineshafts, which examined the distance dependence of the force; and (2) from experiments on the composition dependence of the force. For a reasoned decision to be reached concerning the existence of the Fifth Force, the discords had to be resolved.

For details of this history, see Franklin (1993a).

Tower Gravity Experiments

One way in which the presence of the Fifth Force could be tested was by investigating the distance dependence of the gravitational force, to see if there was a deviation from Newton's inverse-square law. This type of experiment measured the variation of gravity with position, usually in a tower, or in a mineshaft or borehole. All of the experiments used a standard device—a LaCoste-Romberg gravimeter—to measure gravity. The measurements were then compared with the values calculated using a model of the earth, surface-gravity measurements, and Newton's Law of Gravitation.[1] This was a case in which the experiments used the same type of apparatus to measure the same quantity.

Evidence from such measurements had provided some of the initial support for the existence of the Fifth Force. Geophysical measurements during the 1970s and 1980s had given values of G, the universal gravitational constant, that were consistently higher (by about 1%) than that obtained in the laboratory.[2] Because of possible local mass anomalies they were also tantalizingly uncertain.

After the proposal of the Fifth Force, further experimental work was done. At the Moriond workshop in January 1988,[3] Eckhardt presented results from the first of the new tower gravity experiments (Eckhardt et al., 1988).[4] The results differed from the predictions of the inverse-square law by -500 ± 35 μGal (1 μGal $= 10^{-8}$ ms^{-2}) at the top of the tower (Figure 7.1). A second result was also presented at the workshop by the Livermore group (Thomas et al., 1988). They used gravity measurements from five boreholes and found a 2.5% discrepancy between their observed gravity gradient and that predicted by their Newtonian model. This result also differed in magnitude from the 0.52% discrepancy in mineshaft measurements reported by Stacey and in both sign and magnitude from the 0.29% discrepancy reported by Eckhardt. They noted, however, that their measured free-air gradients disagreed with those calculated from their model and concluded "that the model does not reflect the total mass distribution of the earth with sufficient accuracy to make a statement about Newtonian gravity [or about the Fifth Force]" (Thomas et al., 1988, p. 591).

Further evidence for the Fifth Force was provided by a group that measured the variations in gravity in a borehole in the Greenland icecap (Ander et al., 1989). They found an unexplained difference of 3.87 mGal

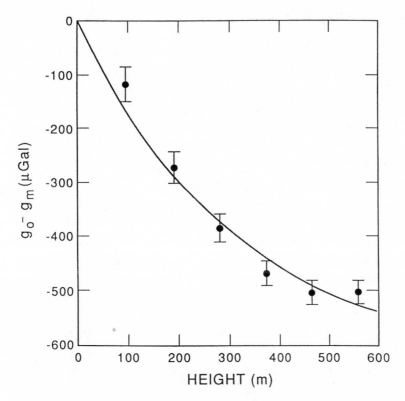

Figure 7.1. Eckhardt et al.'s experimental results fitted to a scalar Yukawa model. The difference between the predictions of Newtonian gravity and the measured values are plotted as a function of the height in the tower. From Fairbank (1988).

between the measurements taken at a depth of 213 m and those taken at 1,673 m. The experimental advantage of the Greenland experiment was the uniform density of the icecap. The disadvantages were the paucity of surface-gravity measurements and the presence of underground geological features that could produce gravitational anomalies.

All of the evidence from tower and mineshaft experiments prior to 1988 supported the Fifth Force. There was, however considerable—although not unambiguous—negative evidence from other types of experiments. Negative evidence from tower experiments would, however, be forthcoming, and it is the discrepancy between the tower results that I address here. (The discord between the other experimental results on the composition dependence of the Fifth Force is addressed in the next section.)

Even before those negative results appeared, questions were raised concerning the positive results. It was not, in fact, the gravity measurements themselves that were questioned. These were all obtained with a standard and reliable instrument. It was, rather, the theoretical calculations used for the theory-experiment comparison that were criticized. One of the important elements needed in these calculations was an adequate model of the earth. Recall that the Livermore group had doubted their own comparison because their model had not given an adequate account of the measured free-air gradients.

The Greenland group's calculation was the first to be criticized. It was subjected to severe criticism, particularly for the paucity of surface-gravity measurements near the location of their experiment (their survey included only 16 such points), and for the inadequacy of their model of the earth. It was pointed out that there were underground features in Greenland of the type that could produce such gravitational anomalies. The Greenland group was criticized for having overlooked plausible sources of error in their experiment-theory comparison and for overlooking plausible alternative explanations of their result. When this result was later presented, the group stated that their result could be interpreted either as evidence for non-Newtonian gravity (a Fifth Force), or explained by local density variations. "We cannot unambiguously attribute it to a breakdown of Newtonian gravity because we have shown that it might be due to unexpected geological features below the ice" (Ander et al., 1989, p. 985).

Parker (a member of the Greenland group) and Bartlett and Tew, suggested that both the positive evidence for the Fifth Force of Eckhardt and collaborators (1988) and that of Stacey and collaborators (1987) could be explained by either local density variations or by inadequate modeling of the local terrain. Bartlett and Tew (1989a) gave more details of their criticism at the 1989 Moriond Workshop. They conceded that it was still an open question as to whether the models of Stacey and Eckhardt properly accounted for local terrain, and presented a calculation arguing that 60–65% of Eckhardt's tower residuals could be explained by local terrain.

Eckhardt disagreed. His group presented a revised value for the deviation from Newtonian gravity at the top of their tower of 350 ± 110 μGal. They attributed this change—a reduction of approximately one-third—to better surface-gravity data and elimination of an elevation bias in their previous survey.[5] "We also had the help of critics who found our claims outrageous" (Eckhardt, 1989, p. 526). They concluded, "nevertheless the experiment and its reanalysis are incomplete and we are not prepared to offer a final result" (p. 526).

The Livermore group presented a result from their gravity measurements at the BREN tower at the Nevada test site (Kasameyer et al., 1989). To overcome the difficulties with their previous calculations, they had extended their gravity survey to include 91 of their own gravity measurements (taken within 2.5 km of the tower) supplemented with 60,000 surface-gravity measurements (taken within 300 km) that were done by others.[6] They presented preliminary results in agreement with Newtonian gravity, reporting that, at the top of the tower, the difference between the measured and predicted values was 93 ± 95 μGal.[7]

Bartlett and Tew continued their work on the effects of local terrain. They argued that the Hilton-mine results of Stacey and his collaborators could also be due to a failure to include local terrain in their theoretical model (Bartlett and Tew, 1989b). They communicated their concerns to Stacey privately. Their view was confirmed when, at the General Relativity and Gravitation Conference in July 1989, Tuck (1989) reported that their group had incorporated a new and more extensive surface gravity survey into their calculation. "Preliminary analysis of these data indicates a regional bias that reduces the anomalous gravity gradient to two-thirds of the value that we had previously reported (with a 50% uncertainty)." With such a large uncertainty, the results of Stacey and his collaborators could no longer be considered as support for the Fifth Force.

Parker and Zumberge (1989), two members of the Greenland group, offered a general criticism of tower experiments. They argued, in some detail, that they could explain the anomalies reported in both Eckhardt's tower experiment and in their own ice-cap experiment using conventional physics and plausible local density variations.[8] They concluded that there was "no compelling evidence for non-Newtonian long-range forces in the three most widely cited geophysical experiments [those of Eckhardt, of Stacey, and their own] . . . and that the case for the failure of Newton's Law could not be established" (p. 31).

The last hurrah for tower gravity experiments that supported the Fifth Force was signaled by Jekeli and collaborators (1990). In this paper, Eckhardt's group presented their final analysis of their data, which

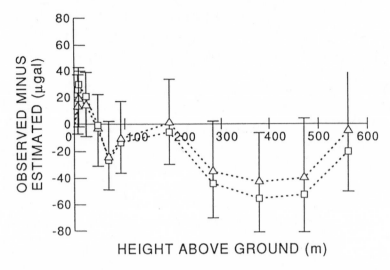

Figure 7.2. Difference between measured and calculated values of *g* as a function of height. No significant difference is seen. From Jekeli et al. (1990).

included a revised theoretical model, and concluded that there was, in fact, no deviation from Newtonian gravity. (See Figure 7.2 and contrast this with their initial positive result, shown in Figure 7.1). Two subsequent tower results also supported Newton's Law (Kammeraad et al., 1990; Speake et al., 1990).

The discord had been resolved. The measurements were correct: It was the comparison between theory and experiment that had led to the discord. It had been shown that the results supporting the Fifth Force could be explained by inadequate theoretical models—failure to account adequately for local terrain or failure to include plausible local density variations. In other words, the Sherlock Holmes strategy had been incorrectly applied. The experimenters had overlooked plausible alternative explanations of the results or possible sources of error.

The careful reader will have noted that it had not been demonstrated that the original theoretical models were incorrect. It had only been shown that the measurements agreed with the theory when plausible sources of error were eliminated. Although this made the positive Fifth Force results very questionable, it was not an airtight argument. The new calculations could have been wrong. Note, however, that the experimenters themselves agreed that the newer models were better.

Scientists make decisions in an evidential context. The Fifth Force was a modification of Newtonian gravity. Newtonian gravity, and its successor, General Relativity, are strongly supported by existing evidence. In addition, there were other credible negative tower gravity results that did not suffer from the same difficulties as did the positive results. There was also, as discussed in the next section, an overwhelming preponderance of evidence against the Fifth Force from other types of experiments. The decision as to which theory-experiment comparison was correct was not made solely on the basis of the experiments and calculations themselves, although one could have justified this. Scientists examined all of the available evidence and came to a reasoned decision about which were the correct results—and concluded that the Fifth Force did not exist.

The Search for a Composition-Dependent Force

The other strand of experimental investigation of the Fifth Force was the search for composition dependence of the gravitational force.[9] The strongest piece of evidence cited when the Fifth Force was originally proposed came from a reanalysis of the Eötvös experiment (Eötvös et al., 1922). The original Eötvös experiment was designed to demonstrate the equality of gravitational and inertial mass for all substances. Eötvös reported equality to about one part in one million. Fischbach and collaborators (1986) had reanalyzed the Eötvös data and reported a large and surprising composition-dependent effect (Figure 7.3).

This was the effect that was subsequently investigated. Two types of composition-dependence experiments are shown in Figure 7.4. To observe the effect of a short-range force such as the Fifth Force, one needs a local mass asymmetry. This asymmetry was provided by either a terrestrial source—a hillside or cliff—or a large, local, laboratory mass. If there were a composition-dependent, short-range force, the torsion pendulum shown in Figure 7.4 would twist in response to it. A variant of this experiment was the float experiment, in which an object floated in a fluid and the difference in gravitational force on the float and on the fluid would be detected by the motion of the float. These experiments were carried out using terrestrial sources.

The results of the first tests for a composition-dependent force appeared in January 1987, a year after the Fifth Force was first proposed. They disagreed. Thieberger (1987), using a float experiment, found

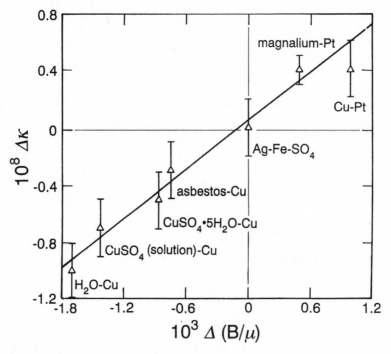

Figure 7.3. $\Delta\kappa$, the fractional change in g, as a function of $\Delta(B/\mu)$. A substantial composition dependence is clearly seen. From Fischbach et al. (1986).

results consistent with the presence of such a force. A group at the University of Washington, headed by Eric Adelberger, and whimsically named the Eöt-Wash group, found no evidence for such a force and set rather stringent limits on its presence (Adelberger et al., 1987).

The results of Thieberger's experiment, done on the Palisades cliff in New Jersey, are shown in Figure 7.5. One can see that the float moves quite consistently and steadily away from the cliff (the y-direction), as one would expect if there were a Fifth Force. Thieberger eliminated other possible causes for the observed motions. These included magnetic effects, thermal gradients, and leveling errors. He also rotated his apparatus by 90° to check for possible instrumental asymmetries, and obtained the same positive result. In addition, he performed the same experiment at another location, one without a local mass asymmetry or cliff, and found no effect, as expected. He concluded:

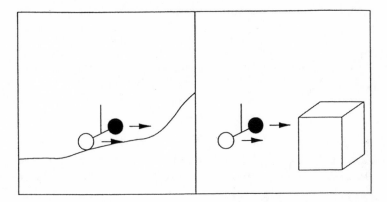

Figure 7.4. Two types of composition-dependence experiments used to search for the Fifth Force. From Stubbs (1990).

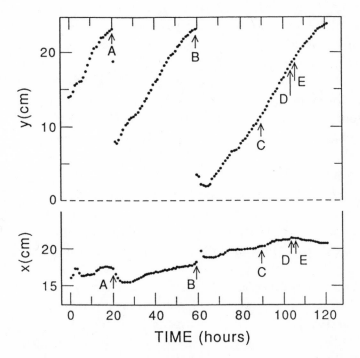

Figure 7.5. Position of the center of the sphere as a function of time. The *y* axis points away from the cliff. From Thieberger (1987).

The present results are compatible with the existence of a medium-range, sub-stance-dependent force which is more repulsive (or less attractive) for Cu than for H_2O. . . . Much work remains before the existence of a new substance-dependent force is conclusively demonstrated and its properties fully character-ized. (Thieberger 1987, p. 1068)

The Eöt-Wash experiment used a torsion pendulum, shown sche-matically in Figure 7.4. It was located on the side of a hill on the University of Washington campus. If the hill attracted the copper and beryllium test bodies used in the apparatus differently, then the torsion balance would experience a net torque. None was observed (Figure 7.6). The group min-imized asymmetries that might produce a spurious effect by machining the test bodies to be identical to within very small tolerances. The test bod-ies were coated with gold to minimize electrostatic forces. Magnetic, ther-mal, leveling, and gravity-gradient effects were shown to be negligible.

The discordant results were an obvious problem for the physics com-munity. Both experiments appeared to be carefully done, with all plausi-ble and significant sources of possible error and background adequately accounted for. Yet the two experiments disagreed.[10]

In this case we are dealing with attempts to observe and measure the same quantity—a composition-dependent force—with very different apparatuses, a float experiment and a torsion pendulum. Was there some unknown but crucial background in one of the experiments that pro-duced the wrong result? To this day, no one has found an error in Thie-berger's experiment, but the consensus is that the Eöt-Wash group is cor-rect and that Thieberger is wrong—that there is no Fifth Force. How was the discord resolved?

In this episode, it was resolved by an overwhelming preponderance of evidence. The torsion pendulum experiments were repeated by others, including Fitch, Cowsik, Bennett, Newman, and again by Eöt-Wash (for details and references, see Franklin [1993a]). These repetitions, in differ-ent locations and with different substances, gave consistently negative results. There was also evidence against the Fifth Force from modern ver-sions of Galileo's Leaning Tower of Pisa experiment performed by Kuroda and Mio (1989) and by Faller (Niebauer et al., 1987). For a graph-ical illustration of how the evidence concerning a composition-dependent force changed with time, see Figures 7.3, 7.7, and 7.8. As more evidence was provided, the initial, and startling, effect claimed by Fischbach and

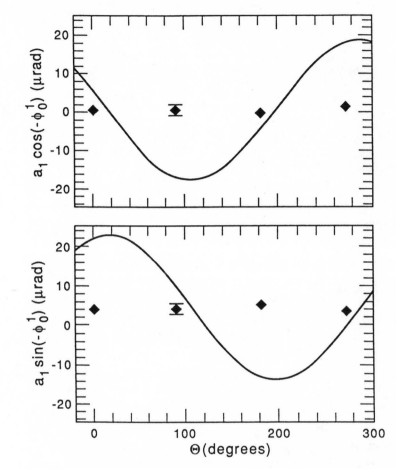

Figure 7.6. Deflection signal as a function of θ. The theoretical curves correspond to the signal expected for α = 0.01 and λ = 100 m. From Raab (1987).

collaborators became far less noticeable. In addition, Bizzeti and collaborators (1988, 1989), using a float apparatus similar to that used by Thieberger, also obtained results showing no evidence of a Fifth Force. (Compare Bizzeti et al.'s results [Figure 7.9] with those of Thieberger [Figure 7.5]). Bizzeti and collaborators' result was quite important. Had they agreed with Thieberger, then one might well have wondered whether there was some systematic difference between torsion-balance experiments and float experiments that gave rise to the conflicting results. But this did not happen. There was, instead, an overwhelming preponderance

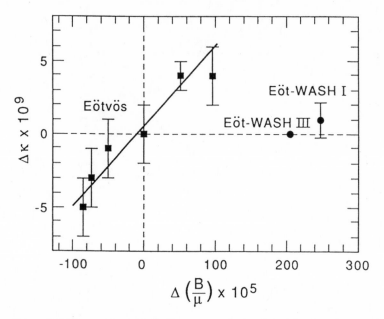

Figure 7.7. Comparison of the Eötvös reanalysis of Fischbach and others (1986) with the results of the Eöt-Wash I and III experiments. The error bar on the Eöt-Wash III datum is smaller than the dot. From Adelberger (1989).

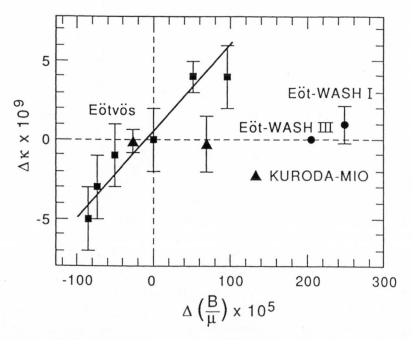

Figure 7.8. The data of Kuroda and Mio (1989) added to Figure 7.7.

176

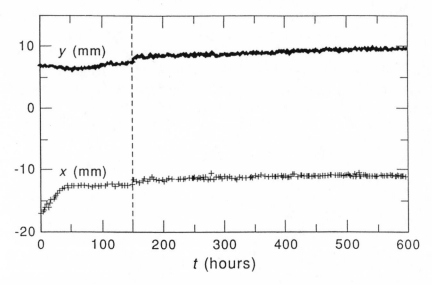

Figure 7.9. Position of the sphere completely immersed in liquid as a function of time. The vertical dashed line marks the time at which restraining wires were removed. From Bizzeti et al. (1988).

of evidence against composition-dependence of the Fifth Force. Even Thieberger (1989) agreed, although he had not found any error in his own experiment:

Unanticipated spurious effects can easily appear when a new method is used for the first time to detect a weak signal. . . . Even though the sites and the substances vary, effects of the magnitude expected have not been observed. . . . It now seems likely that some other spurious effect may have caused the motion observed at the Palisades cliff.[11] (p. 810)

In both instances discussed in this chapter—the composition dependence and the distance dependence of the proposed Fifth Force—the decision that such a force did not exist was made on the basis of reasons that allow us to consider experimental results as the basis for scientific knowledge. In the case of the distance dependence, it was shown that the positive results were obtained by overlooking effects in the theoretical calculations that resulted in an incorrect experiment-theory comparison. This, combined with credible negative results, argued against the existence of the Fifth Force. The discrepancy between the Thieberger and

Adelberger results on the composition dependence of the Fifth Force was resolved by an overwhelming preponderance of evidence. In addition, Bizzeti and collaborators, using an apparatus quite similar to that of Thieberger, found no evidence for the Fifth Force. This argued against any crucial difference between the different types of apparatus being responsible for the discordant results.

8

William Wilson and the Absorption of β Rays

In the first decade of the twentieth century, physicists believed that the β particles emitted in radioactive decay were monoenergetic and that such monoenergetic electrons would be absorbed exponentially in passing through matter.[1] Conversely, they also believed that if electrons followed an exponential absorption law, then they were monoenergetic. There was evidence supporting this view. William Wilson, however, with some supporting evidence from other experimentalists, showed conclusively that this view was wrong. Within a very short period of time, the physics community accepted his results. He also showed that the previous experimental results, on which the view of exponential absorption had been based, were, in fact, correct. They had been misinterpreted. This was an episode in which the discord was apparent, not real. Nevertheless, because the participants believed the discord was real, the method by which the discord was resolved is relevant to our discussion.

The Exponential Absorption of β Rays

In 1902, Kaufmann (1902) had demonstrated that radium emitted electrons with a wide range of velocities. A similar result also was found by

For a more detailed discussion, see Franklin (2002).

Meyer and von Schweidler (1899) and Becquerel (1900). Despite the evidence provided, the physics community did not accept, at this time (the first decade of the twentieth century), that the energy spectrum of electrons emitted in β decay was continuous. There were plausible reasons for this. Physicists argued that the sources used by both Kaufmann and Becquerel were not pure β-ray sources, but contained several elements, each of which could emit electrons with different energies. In addition, even if the electrons were initially monoenergetic, each electron might lose a different amount of energy in escaping from the radioactive source. This view was due, in part, to a faulty analogy with α decay. Bragg (1904) had argued earlier that each of the α particles emitted in a particular decay has the same, unique energy, as well as a definite range in matter. Physicists at the time thought, by analogy with the α particles, that the β rays would also be emitted with a unique energy. Physicists also knew that electrons did not have a unique range in matter.

The difference between the behavior of the α and β particles was due to the difference in their interactions with matter. Alpha particles lose energy almost solely by ionization, whereas electrons lose energy by several processes, including ionization, scattering, and processes unknown to physicists in the early twentieth century. It was believed that monoenergetic electrons would follow an exponential absorption law when they passed through matter. This was a reasonable assumption for the physicists of that time: If electron absorption was dominated by the scattering of electrons out of the beam, and if the scattering probability per unit length was constant, then an exponential absorption law would follow. As Bragg (1904) stated:

Nevertheless it is clear that β rays are liable to deflexion through close encounters with the electrons of atoms; and therefore the distance to which any given electron is likely to penetrate before it encounters a serious deflexion is a matter of chance. This, of course, brings in an exponential law. (p. 720)

Early experimental work on electron absorption gave support to such an exponential law and therefore to the homogeneous (monoenergetic) nature of β rays, particularly the work of Schmidt (1906, 1907). Schmidt fitted his absorption data for electrons emitted from different radioactive substances with a single exponential or a superposition of a few exponentials. Figure 8.1 shows the absorption curves that Schmidt obtained for electrons from radium B and from radium C.[2] The logarithm of the ionization (a measure of the electron intensity) decreases linearly with

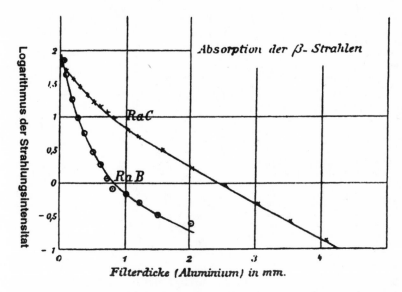

Figure 8.1. Schmidt's result on the absorption of β rays. The logarithm of the electron intensity (ionization) is plotted as a function of absorber thickness. Each of the curves is a reasonable fit to two straight lines, indicating to Schmidt both exponential absorption and the presence of two groups of monoenergetic electrons for each substance. From Schmidt (1906).

the thickness of the absorber, which indicates an exponential absorption law. Each curve actually consists of two straight line segments, showing the superposition of two exponentials. Schmidt (1906) interpreted this result as demonstrating that two groups of β rays were emitted in each of these decays, each with its own unique energy and absorption rate:

We have seen that the β-rays from radium are absorbed according to a pure exponential law within certain filter thicknesses. Should this not be taken to mean that there exists a certain group [of rays] with a constant absorption coefficient among the totality of β-radiations? Indeed, could we not go one step further and interpret the total action of β-rays in terms of a few β-ray groups [each] with a constant absorption coefficient? (Translated in Pais 1986, p. 149)

There was, in fact, a circularity in the argument. If the β rays were monoenergetic, then they would give rise to an exponential absorption law. If they followed an exponential absorption law, then they were monoenergetic. As Rutherford (1913) remarked:

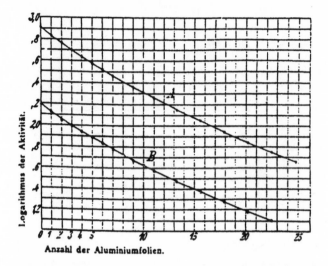

Anzahl der Aluminiumfolien.

Figure 8.2. The β-ray absorption curve obtained by Hahn and Meitner for mesothorium. The curves are a reasonable fit to straight lines. From Hahn and Meitner (1908a).

Since Lenard had shown that cathode rays . . . are absorbed according to an exponential law, it was natural at first to assume that the exponential law was an indication that the β rays were *homogeneous,* i.e. consisted of β particles projected with the same speed. On this view, β particles emitted from uranium which gave a nearly exponential law of absorption, were supposed to be homogeneous. On the other hand, the β rays from radium which did not give an exponential law of absorption were known from other evidence to be heterogeneous. (pp. 209–10)

This association of homogenous electrons with an exponential absorption law informed early work on the energy spectrum in β decay. This was the situation in 1907, when Lise Meitner, Otto Hahn, and Otto von Baeyer began their work on the related problems of the absorption of electrons in matter and of the energy spectrum of electrons emitted in β decay.[3] They first examined the absorption of electrons emitted in the β decay of several complex substances: uranium + uranium X (^{234}Th), radiolead + radium E, radium E alone, and radium. They found that the absorption of these electrons did, in fact, follow an exponential law, confirming the results obtained by Schmidt (Figure 8.2). They formulated the simple and attractive hypothesis that each pure element emitted a single group of monoenergetic β rays. The multielement sources they used yielded absorption curves that consisted of several superposed exponentials. The only exception seemed to be mesothorium-2. As Hahn (1966) later remarked, "but we felt so certain about the uniformity of beta rays from uniform elements that we explained the noncompliance of mesothorium-2 by a still not understood complexity in the nature of mesothorium-2" (pp. 53–54).

The Experiments of William Wilson

The evidential situation changed dramatically with the work of William Wilson. Wilson investigated what was, in retrospect, a glaring omission in the existing experimental program—the actual investigation of the velocity dependence of electron absorption (Wilson, 1909). He noted that his "present work was undertaken with a view to establishing, *if possible*, the connection between the absorption and velocity of β rays. *So far no actual experiments have been performed on this subject*" (p. 612, emphasis added). Although Schmidt's experiments had provided some information on the subject, there had been no real investigation of the issue. Wilson commented that "It has generally been assumed that a beam of homogeneous rays is absorbed according to an exponential law, and the fact that this law holds for the rays from uranium X, actinium, and radium E has been taken as a criterion of their homogeneity" (p. 612). Wilson questioned that assumption:

The assumption is open to many objections, for the exponential law may be due to rays of different types being mixed in certain proportions. If the distribution of the rays and their velocity do not change in passing through matter, and if the absorption of the particles is proportional to the number present, we should expect an exponential law of absorption [as previous experimenters had assumed], but *if their speed diminishes*, the absorption should be greater the greater the thickness of matter traversed. (p. 612, emphasis added)

Wilson included not only detailed arguments for the credibility of his result, but also gave careful consideration to backgrounds that might mask or mimic the effect he wished to measure and how he dealt with them. He also included an explanation of why his results differed from those obtained previously by other experimenters.

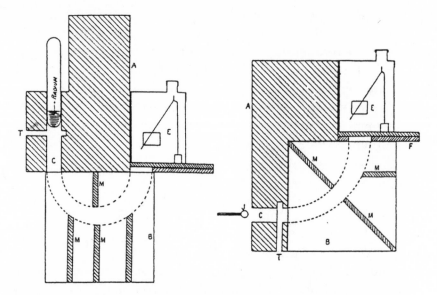

Figure 8.3. W. Wilson's experimental apparatuses for measuring the absorption of β rays. Electrons from the radioactive sources pass through slit C, are bent by a magnetic field perpendicular to the plane of the paper, and pass through the slits in plate MM and F. This defines a range of radii of curvature and thus, a range in momentum or velocity. Varying amounts of absorber were placed above slit F and the β rays were detected by the electroscope. From Wilson (1909).

Wilson used radium as the source of his electrons. He noted that Kaufmann had shown that radium emitted electrons with a wide range of velocities. Wilson selected electrons within a narrow band of velocities—an almost monoenergetic beam—and investigated their absorption. He stated his remarkable conclusion at the beginning of his paper: "Without entering at present into further details, it can be stated that the ionisation [the electron intensity] did not vary exponentially with the thickness of matter traversed. But, except for a small portion at the end of the curve, followed approximately a linear law" (p. 613). This result contradicted those of Schmidt, Meitner, Hahn, and von Baeyer.

The two different versions of Wilson's experimental apparatus are shown in Figure 8.3. In the apparatus on the left, a radium-bromide source was placed at C. The collimated β rays from the decay of radium were bent in a circular path by a magnetic field perpendicular to the plane of the paper. The rays passed first through slits MM and F, then through an absorber, and were detected by the ionization produced in electroscope E. The radius of the circular path is proportional to the velocity of the electrons, so that by selecting only electrons with certain path radii, Wilson was selecting electrons within a certain velocity (or energy) range, whose width was approximately 10%. Varying the strength of the magnetic field changed the velocity of the selected electrons, so that the absorption of the electrons as a function of velocity could be measured. Initially Wilson found that most of the electrons emitted were absorbed before leaving the radium-bromide source. To increase the signal, in later experiments he substituted a thin-walled glass bulb containing radium emanation (radon, a radioactive gas emitted by radium) for the original radium-bromide source.

There were important sources of background, however, that limited the accuracy of the measurement. Wilson devoted considerable care and effort to reducing this background, and in cases where it could not be eliminated, to measuring the size of the background signal so that it could be subtracted from the total signal to obtain a correct measurement. A major source of such background were the γ rays that were also emitted by the radioactive source. These γ rays produced ionization in the electroscope that mimicked that produced by the decay electrons. This background effect was typically about 60% of the entire ionization produced, and for thick absorbers, when the number of decay electrons remaining was greatly reduced, accounted for almost all of the ionization produced. If the background could not be eliminated or greatly reduced, then the experiment would be impossible. Wilson replaced the radium-bromide source with one consisting of radium emanation (radon) and reduced the γ-ray background to <20% of the total signal. He also measured the ionization produced by the γ rays by inserting a lead plate at slot T (Figure 8.3). The plate was thick enough to eliminate all of the decay electrons, but left the γ-ray background essentially unchanged. The remaining ionization measured by the electroscope then was entirely due to the γ-ray background, which was measured and subtracted from the total signal for each setting. Background from electrons scattering from other parts of the apparatus was greatly reduced by the lead screens (M and MM in Figure 8.3).

Wilson's results are shown in Figure 8.4. The upper graph shows the ionization (not its logarithm) for various velocities as a function of absorber thickness. It is clearly linear, and not exponential. This is made clear in the lower graph, in which the logarithm of the ionization is plotted against absorber thickness. As we have seen earlier in the results obtained by Schmidt, if the law of absorption were exponential, then this graph would be a straight line. It is not.

Wilson recognized that his result, which disagreed with all of those obtained previously, needed to be defended carefully. He identified three possible influences that might affect the absorption curves and give an incorrect result: (1) the lack of saturation in the ionization current; (2) the shape and size of the electroscope opening; and (3) the proximity of the magnetic field to the electroscope, which might cause irregularities in its operation. Wilson compared the time it took for the gold leaf of the electroscope to traverse very different parts of the scale for various values of the ionization. If the ionization was saturated, then the ratio of the times should be constant when the ionization level was varied, and it was. This result was further checked by measuring the absorption curves obtained using two sources of very different strengths. They were identical, further indicating that saturation was not a problem.

Wilson also calibrated his electroscope by showing that it gave the same absorption curve for actinium as that obtained in previous measurements. He further checked for possible magnetic-field effects by measuring that same absorption curve with the magnetic field on and with it off. No difference was observed, indicating that the magnetic field did not affect the operation of the electroscope or his result. Wilson's results were internally consistent. He obtained the same result with both versions of his experimental apparatuses, even though the magnetic field required to deflect the electrons into the electroscope was far larger in the first device than in the second.

Wilson's results were credible. He had either reduced the background effects or measured them so that they could be subtracted. He had also shown that none of the effects that might have compromised his results were present. He had calibrated his apparatus and obtained independent confirmation of his result using two different experimental apparatuses. He had eliminated plausible alternative explanations of his result, and was left with the conclusion that the result was correct.

How could such capable physicists as Wilson, Schmidt, and the trio of Hahn, Meitner, and von Baeyer reach such different conclusions about electron absorption? Wilson had shown that the absorption of monoenergetic electrons was approximately linear, whereas the others had

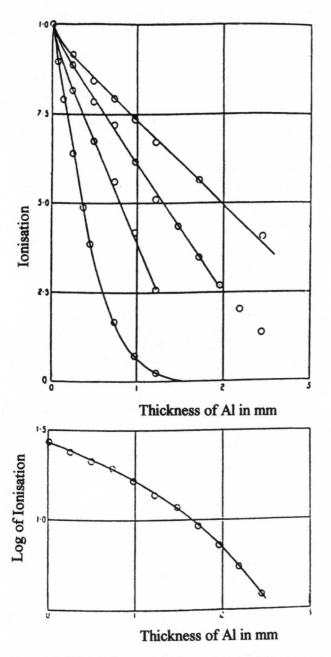

Figure 8.4. Wilson's β-ray absorption curves. (top) The ionization produced (or electron intensity), not its logarithm, is plotted as a function of absorber thickness for different velocities. The curves are quite reasonable fits to straight lines, except perhaps near their end, indicating a linear, rather than an exponential, absorption law. (bottom) The logarithm of the intensity as a function of absorber thickness. It is not a straight line, as would be expected for exponential absorption. From Wilson (1909).

found that electron absorption followed an exponential law. In retrospect, the simple explanation is that Wilson had actually measured the absorption of monoenergetic electrons in groups that had various velocities, whereas the others had assumed that they were measuring the absorption of monoenergetic electrons when they were, in fact, measuring the absorption of electrons with a continuous energy spectrum. What makes Wilson's paper so fascinating is that he provided an explanation for these conflicting results. The other experimental results were not incorrect; they had been misinterpreted.

Wilson (1909) devoted a section of his paper to an "Explanation of the Exponential Law found by various Observers for the Absorption of Rays from Radio-Active Substances." He began:

Before entering into a discussion as to the meaning of the absorption curves obtained, it is preferable to try to explain why various observers have found that the rays from Uranium X, radium E, and actinium are absorbed according to an exponential law with the thickness of matter traversed. The fact that homogeneous rays are not absorbed according to an exponential law suggests that *the rays from these substances are heterogeneous.* (pp. 621–22, emphasis added)

Wilson then provided an explanation. He began with data from Schmidt's work that showed the ionization produced as a function of the velocity of the emitted rays. Schmidt had found a range of such velocities, but had not interpreted that result as indicating that the primary electrons were heterogeneous. He and others believed that they were emitted with a unique energy, but that they then lost energy by some unknown process. Wilson showed that the ionization curve produced varied with the amount of matter through which the electrons had passed (Figure 8.5).[4] The figure shows the electron intensity as a function of momentum. Curves *a, b,* and *c* were obtained with thicknesses of aluminum of 0, 0.489, and 1.219 mm, respectively. Not only was the total ionization reduced, but the lower-velocity electrons were completely absorbed when the absorber thickness was increased. He calculated this effect for various absorber thicknesses and found that the total ionization produced by such heterogeneous electrons as a function of that thickness indeed did follow an exponential law (Figure 8.6). He concluded that "It is thus clear that the exponential curve for the absorption of rays is not, as has been widely assumed, a test of their homogeneity, but that in order that the exponential law of absorption should hold, we require a mixture of rays of different types" (pp. 623–24).

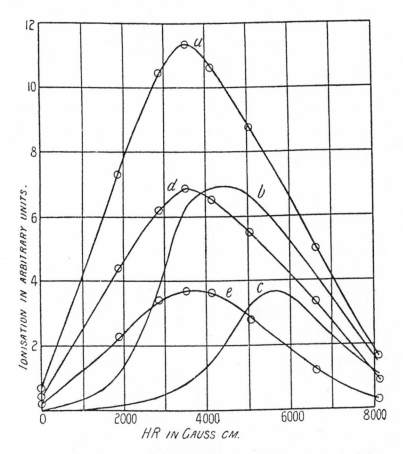

Figure 8.5. Ionization as a function of momentum for different thicknesses of absorber. Curves *a, b,* and *c* are for aluminum absorbers of thickness 0, 0.489, and 1.219 mm, respectively, placed just under the electroscope in Figure 8.3. Curves *d* and *e* are for thicknesses of 0.489, and 1.219 mm, respectively, placed at T (figure 8.3, left side of right-hand apparatus), just before entering the magnetic field. If the β rays lose velocity in passing through matter, then curves *d* and *e* should be shifted to the left. They are. From Wilson (1909).

There was, however, existing evidence that disagreed with Wilson's result that the velocity of electrons diminished as they passed through matter. Schmidt (1907) had used the apparatus shown in Figure 8.7 to investigate the constancy of the β-particle velocity. The β rays from a radium E source at A were bent by a magnetic field perpendicular to the plane of the paper so that they passed through a semicircular canal ABC

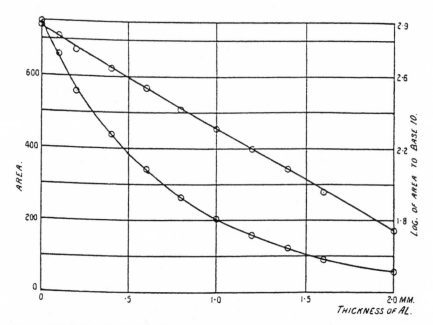

Figure 8.6. Wilson's calculated absorption curve assuming an inhomogeneous energy spectrum for the emitted β rays. Wilson calculated values of the intensity as a function of absorber thickness using an initially inhomogeneous beam of electrons. The lower curve plots the intensity, whereas the upper curve is the logarithm of the intensity. The absorption is predicted to be linear, not exponential. From Wilson (1909).

and then passed into an ionization chamber. Schmidt adjusted the field strength to a value H_0, which resulted in the maximum ionization (i.e., the maximum number of β rays). He then placed aluminum foils between the radioactive source and the canal entrance. The β rays passed through the absorber. Once again he adjusted the field strength to obtain the maximum number of β particles. If the velocity had not changed in passing through the absorber, then that value would be H_0. It was. This certainly cast doubt on Wilson's result that the β particles lost energy in passing through matter, a result he needed to explain why he had observed linear absorption whereas others had found an exponential absorption law.

Wilson showed that although Schmidt's experimental result was correct, his interpretation of that result was incorrect. Wilson measured the absorption of β rays as a function of momentum for various absorber thicknesses. The measurements were made under two different sets of

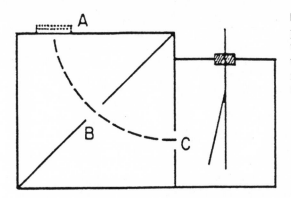

Figure 8.7. Schmidt's apparatus for demonstrating that β rays do not change velocity in passing through matter. From Heilbron (1967).

conditions. For curves *a*, *b*, and *c* in Figure 8.5, the absorber (thickness 0, 0.489 mm, and 1.219 mm, respectively) was placed just under the electroscope (see Figure 8.3, right side). For curves *d* and *e* (0.489 mm and 1.219 mm) the absorber was placed at T, before the electrons entered the magnetic field. If electrons do not change velocity in passing through matter, curves *d* and *e* should be identical to curves *b* and *c*. They clearly are not; the shift of the curves to the left demonstrates that electrons lose velocity (energy) in passing through matter. Once again Wilson had provided an explanation of an incorrect interpretation of an experimental result.

Wilson's results also explain why the experiments of Schmidt apparently show no change in the velocity of the rays. According to the views expressed in Wilson's paper, he was dealing with heterogeneous rays and the position of the maximum should therefore move to the higher fields if the velocity of the rays does not change. The actual decrease in velocity, however, brings the maximum point back to practically the same position as before (Wilson, 1909).

Initial Reaction of the Physics Community

Wilson's negative result on the exponential absorption of β rays received support from further work by Schmidt (1909). Schmidt inferred from his data that electrons did change their velocity in passing through matter, confirming Wilson's result, and he also found that the electrons were not always absorbed exponentially. He did not mention or cite Wilson's results, however. This was not the case in the paper by Hahn and Meitner (1909b), published in the same journal issue as Schmidt's paper. Hahn

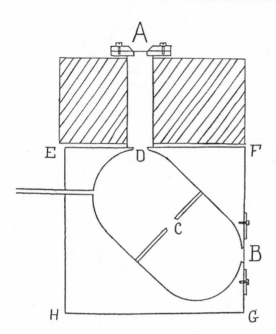

Figure 8.8. One of Crowther's chambers for investigating whether electrons lost velocity in passing through matter. Two chambers were placed so that the window A of one chamber was opposite window B of the other. There were separately adjustable magnetic fields perpendicular to the plane of the paper in each chamber. From Crowther (1910).

and Meitner argued that Wilson's results showed rather that the β-decay electrons were monoenergetic and that they did not lose energy in passing through matter. They also suggested that the energy spread in Wilson's electron beam was too large. Wilson (1910b) responded and argued persuasively for the homogeneity of his electron beam. He noted that his differences with Hahn and Meitner did not concern the correctness of their respective experimental results, but rather the interpretation of those results. Hahn and Meitner performed no further experiments on electron absorption, and in a later account, Meitner (1964) remarked that they had realized that to say anything about the velocity of the electrons, they had to use deflection in a magnetic field, just as Wilson had done.

The issue of whether the electrons emitted in β decay were monoenergetic or had a continuous energy spectrum would not be resolved until the experimental work of Ellis and Wooster (1927).[5] The question of whether electrons lose energy in passing through matter was, however, immediately investigated and resolved in favor of Wilson. Crowther (1910) soon reported results on that very question. He noted that there already existed a considerable amount of indirect evidence on the sub-

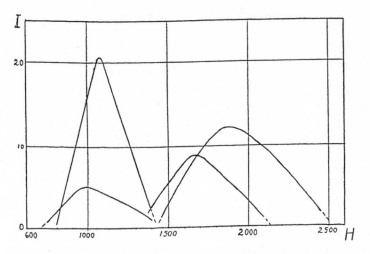

Figure 8.9. Electron intensity as a function of magnetic field for two different initial electron velocities. The two curves on the right correspond to a high initial velocity. The upper and lower of these curves were obtained with 0.0 and 0.47 mm of aluminum absorber, respectively. Not only is the intensity reduced, but the lower curve is shifted to a lower magnetic field value, demonstrating that the electrons have lost velocity in passing through the absorber (similarly for the curves for the initially lower velocity electrons on the left). From Crowther (1910).

ject, citing the work of both Wilson and Schmidt. Crowther proposed to investigate the question directly by measuring the velocity of electrons before and after they passed through an absorbing layer. His apparatus is shown in Figure 8.8. Radium was placed at either A or B (depending on whether a parallel beam of electrons was needed), and the entire apparatus placed in a magnetic field:

Two chambers were made and placed so that the window A of the one came directly opposite the window B of the other. The two magnetic fields [each apparatus had its own, separately adjustable magnetic field] were arranged so that the rays of the proper velocity would be deflected round the two systems and emerge finally into an ionization chamber of the usual pattern. (p. 446)

Fig. [8.9] shows the effect of interposing a sheet of aluminum 0.47 mm in thickness between the two systems, for two different velocities of the incident beam. The upper curve in each case is the curve obtained for the incident beam in the absence of the absorbing sheet. The ordinates represent the intensity of the

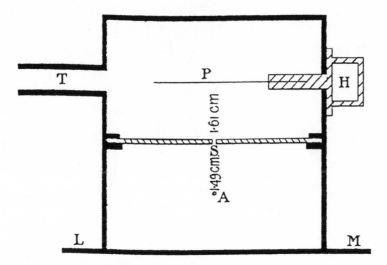

Figure 8.10. Gray's apparatus for investigating the absorption of β rays. Electrons from the radioactive source A passed through the slit S and struck a photographic plate at P. The entire apparatus was placed in a magnetic field so that electrons of different velocities would follow different paths. From Gray (1910).

rays passing through the two systems, as measured by the ionization produced; the abscissae measure the magnetic field acting upon the second system.

It will be seen that in each case the introduction of the absorbing sheet produces a very definite displacement of the curve in the direction of smaller velocities. . . .

It is evident therefore that there is a small, but perceptible decrease in the velocity of the β-rays as they pass through absorbing media. (p. 448)

Crowther had shown that electrons lost energy in passing through matter. His results supported those of Wilson. Further support was provided by the experiments of J. A. Gray, who, along with Wilson, was working in Manchester with Ernest Rutherford. To avoid difficulties arising from the decay of several elements in the same source, Gray used Radium E (^{214}Bi), a single-element source. His apparatus is shown in Figure 8.10. Electrons from the radioactive source A passed through the slit S and struck a photographic plate at P. The entire apparatus was placed in a magnetic field, so that electrons of different velocities would follow different paths. Gray's results (1910) are shown in Figure 8.11:

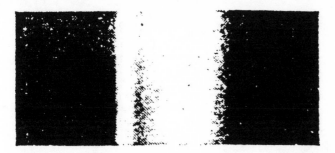

Figure 8.11. Gray's photograph of the intensity of electrons of different velocities. No evidence of a single energy or of a line spectrum is seen. "The narrow band [on the left] is caused by the undeflected rays in the absence of a magnetic field. The other band [on the right], or magnetic spectrum as it may be called, shows no sign of bands, the spectrum being quite continuous." From Gray (1910).

There was no sign of a set or sets of homogeneous β-rays. The narrow band is caused by the undeflected rays in the absence of a magnetic field. The other band, or magnetic spectrum as it may be called, shows no sign of bands, the spectrum being quite continuous. (p. 138)

The results also show a broad spectrum of electron velocities. Gray measured the absorption of these electrons, and his results are shown in Figure 8.12. The logarithm of the intensity as a function of the thickness of the aluminum absorber "is practically a straight line," indicating exponential absorption. "[We] see that β-rays, which are very nearly absorbed according to an exponential law, are by no means homogeneous" (p. 140). This conclusion conformed to Wilson's view on exponential absorption.

Surprisingly, Gray did not emphasize the continuous energy spectrum of the electrons emitted in β decay that seems to be indicated by his results. This may have been due, in part, to his focus on electron absorption (Wilson had been similarly focused). Gray (1910) concluded:

Summing up the experiments, which as we have seen, confirm Wilson's results, we may say that—

1. β-rays, which are absorbed according to an exponential law, are not homogeneous.

2. β-rays must fall in velocity in traversing matter. (p. 141)

In a companion paper, Wilson (1910a) presented further evidence to support his view that electrons lost velocity in passing through matter.

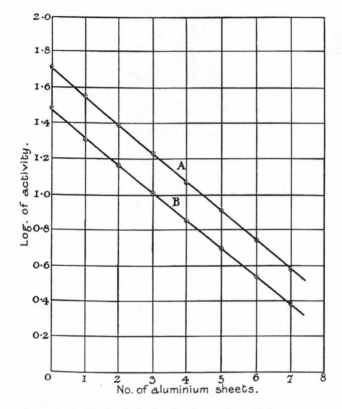

Figure 8.12. Gray's results for the absorption of inhomogeneous β rays, for two different experimental conditions. The logarithm of the intensity of electrons as a function of absorber thickness is plotted as a function of absorber thickness. Both curves are straight lines, indicating an exponential absorption law. From Gray (1910).

His experimental apparatus, similar to that used by Crowther, is shown in Figure 8.13. Electrons emitted by a radioactive source A were bent in a circular path by a uniform magnetic field C. If the electrons followed a circular path of a certain radius, they would pass through hole O. The electrons that passed through hole O were then bent by a uniform magnetic field D and detected in electroscope E. Absorbers of various thicknesses were placed at Q, and the velocity change, if any, measured.[6] Wilson described the experiment as follows:

By passing currents of known strength though the electromagnets C and D, approximately homogeneous radiation was allowed to pass through the hole O

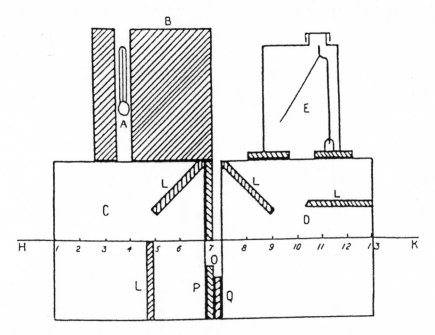

Figure 8.13. Wilson's apparatus for investigating whether electrons lose velocity in passing through matter. It is similar to that used by Crowther (Figure 8.8). A radioactive source is placed at A and the emitted electrons bent in a circular path by separately adjustable magnetic fields in each chamber. Absorbers of varying thickness were placed at Q. From Wilson (1910a).

into the magnetic field D. The field in D was varied, while that in C was kept constant, so that the same bundle of approximately homogeneous rays passed through the hole O during the whole of the experiment. The ionisation in the electroscope was determined for each value of the field in D, and the values thus obtained were plotted against the current in D. The rays were then made to pass through various sheets of aluminum placed in the slot in Q before they entered the second magnetic field and the experiments repeated. (p. 144)

Wilson's results for various absorber thicknesses are shown in Figure 8.14. "It will be noticed that these maximum points move to the lower fields as the sheets of aluminum are interposed in the path, *proving conclusively that the velocity of the rays decreases by an appreciable amount as they pass through matter*" (p. 145).

Meitner, Hahn, and von Baeyer made improvements in their experimental apparatus to increase the energy resolution in their β-ray spectra

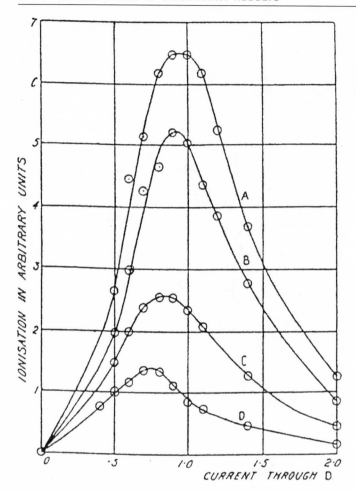

Figure 8.14. Wilson's results on whether electrons lose velocity in passing through matter. The intensity of electrons is plotted as a function of the current in the magnet in chamber D (Figure 8.13), after the absorber, for various absorber thicknesses and for different initial velocities. As the absorber thickness is increased (curves A–D), the intensity is reduced and the peak is shifted to lower fields, demonstrating that the electrons have lost velocity in passing through the absorber and that the loss is larger for thicker absorbers. From Wilson (1910a).

(Figure 8.15). Electrons emitted from the radioactive source S were bent in a magnetic field, passed through a small slot F, and then struck a photographic plate P. Electrons of the same energy would follow the same path and produce a single line on the photographic plate. The results showed a line spectrum and still seemed to support the view that there

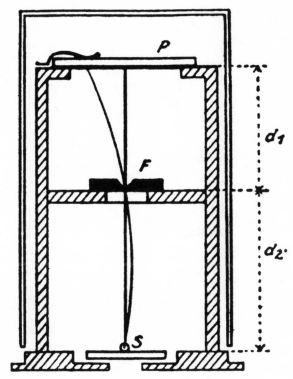

Figure 8.15. The experimental apparatus used by Meitner, Hahn, and von Baeyer. The β rays emitted by the source S are bent by a magnetic field, pass through a slit at F, and strike the photographic plate P. From Hahn (1966).

was one unique value for the electron energy for each radioactive element. The best photograph obtained with a thorium source shows two strong lines, corresponding, the experimenters believed, to the β rays from the two radioactive substances present (Figure 8.16). There were some problems, however. There are some weak lines in the photograph that are difficult to explain from the viewpoint of one energy line per element (von Baeyer et al., 1911a):

The present investigation shows that, in the decay of radioactive substances, not only α-rays but also β-rays leave the radioactive atom with a velocity characteristic for the species in question. This lends new support to the hypothesis of Hahn and Meitner. (p. 279)

Further improvements to the apparatus, including stronger and thinner radioactive sources, improved the quality of the photographs obtained, but showed a complexity of electron velocities that made it dif-

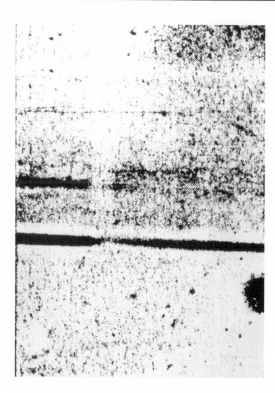

Figure 8.16. The first line spectrum for β decay published by Meitner, Hahn, and von Baeyer. The two observed lines were thought to be produced by the two radioactive elements present in the source. From von Baeyer et al. (1911a).

ficult to argue for the Hahn-Meitner hypothesis (Figure 8.17). As Hahn (1966) later wrote, "Our earlier opinions were beyond salvage. It was impossible to assume a separate substance for each beta line" (p. 57)." Von Baeyer and collaborators (1911b) also conceded that the exponential absorption law "could not be a criterion for the homogeneity of the radiation as Hahn and Meitner, in contrast to other scientists[7] have assumed" (p. 379)."[8]

Gray and Wilson (1910) proved the heterogeneity of electrons emitted from a thick layer of radium E. They remarked on the recent discussions and experimental evidence concerning the exponential absorption of electrons and on the decrease in the velocity of electrons as they pass through matter. They concluded that "It follows as a necessary consequence of these results that β rays which are absorbed exponentially by aluminum are not homogeneous" (p. 870). They noted, however, that recent work by von Baeyer and Hahn had shown "that the β rays from several radioactive products possess a considerable degree of homogene-

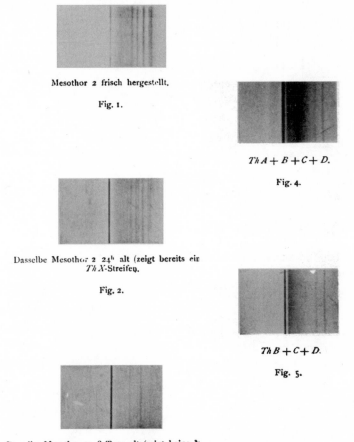

Mesothor 2 frisch hergestellt.

Fig. 1.

$Th A + B + C + D.$

Fig. 4.

Dasselbe Mesothor 2 24ʰ alt (zeigt bereits ein
$Th X$-Streifen.

Fig. 2.

$Th B + C + D.$

Fig. 5.

Dasselbe Mesothor 2; 8 Tage alt (zeigt keine M
thorstreifen mehr, dafür die $Th X$- und $Th A$-Strei

Figure 8.17. The complex β-ray energy spectra obtained by Meitner, Hahn, and von Baeyer with their improved apparatus. A large number of lines are seen. (a) Mesothorium-2, freshly prepared. (b) Mesothorium-2, 24-hours old, showing a faint line from thorium-X. (c) Mesothorium-2, eight days old. The mesothorium lines have disappeared, but lines from thorium-X and thorium-A begin to show. (d) Thorium A + B + C + D. (e) Thorium B + C + D. From Hahn (1966).

ity" (a reference to the line spectra that had been found) and went on to state:

We have no definite evidence so far that the rays from such thin layers as they used are absorbed according to an exponential law. Gray by the same method showed that the β rays from a thick layer of radium E are distinctly hetero-

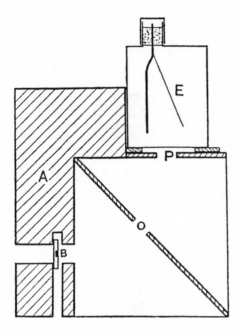

Figure 8.18. The apparatus of Gray and Wilson for investigating the absorption of β rays. From Gray and Wilson (1910).

geneous, although they are absorbed according to an exponential law by aluminum. In view of the experiments of v. Baeyer and Hahn the following experiments were performed. (pp. 870–71)

Surprisingly, they did not mention Gray's own very recent result, discussed above, that the electrons from a thin film of radium E were emitted with widely different velocities.

The experimental apparatus of Gray and Wilson is shown in Figure 8.18. Electrons from a radium-E source at B were bent in a circular orbit of fixed radius by a magnetic field, passed through holes O and P, and were detected by electroscope E. The ionization in the electroscope was measured as a function of magnetic field. The experiment was repeated with aluminum absorbers of various thickness placed just below the electroscope. The results are shown in Figure 8.19. The graph clearly shows that lower-field (lower velocity or energy) electrons are absorbed more easily. "It will be noticed that the rays which produced the maximum ion-

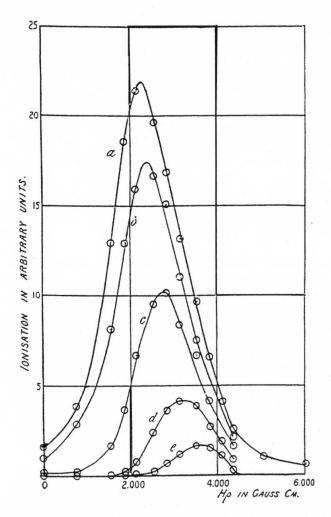

Figure 8.19. The ionization in the electroscope (electron intensity) plotted as a function of electron momentum. The experiment was repeated with aluminum absorbers of various thickness placed just below the electroscope (Figure 8.18). Curves *a, b, c, d,* and *e* were obtained with aluminum thicknesses 0.0, 0.067, 0.245, 0.489, and 0.731 mm, respectively. The graph clearly shows that low momentum (energy) electrons are absorbed more easily. From Gray and Wilson (1910).

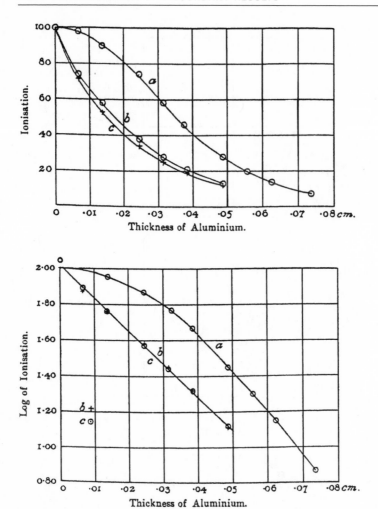

Figure 8.20. Wilson's experimental plots on the absorption of a beam of electrons. (top) Curves *b* and *c* show the intensity of homogeneous electrons as a function of absorber thickness after they have passed through a platinum sheet rendering them inhomogeneous. Curve *a* shows the absorption of the homogenous electrons. (bottom) The logarithm of the ionization is plotted. The curves for *b* and *c* are straight lines, indicating exponential absorption for *inhomogeneous* electrons. Curve *a*, for monoenergetic electrons, is clearly not an exponential. From Wilson (1912).

ization when no aluminum was placed under the electroscope are practically all absorbed by a thickness of .73 mm Al, while for rays corresponding to the higher fields appreciable quantities are still transmitted" (Gray and Wilson, 1910, p. 873).

Gray and Wilson (1910) also measured absorption curves for electrons of different energies directly. They found, once again, that the lower-energy electrons were more easily absorbed and that the absorption for such almost monoenergetic electrons was not exponential:

It has been shown above that from a pencil of β rays which is absorbed according to an exponential law, rays of widely different penetrating powers can be separated out. It follows, therefore, that absorption of β rays according to an exponential law is no criterion of homogeneity. (p. 875)

By 1911, Hahn, Meitner, von Baeyer, Gray, Wilson, Crowther, and, no doubt, everyone else in the physics community were in agreement. Monoenergetic electrons were not absorbed exponentially and exponential absorption was not an indication that they were monoenergetic.

The *coup de grace* was administered by Wilson. Wilson was not satisfied with a mere calculation to show that other experimenters had misinterpreted their results on electron absorption. In subsequent experimental work, he showed that an inhomogeneous beam of electrons was absorbed exponentially (Wilson, 1912). He began with a monoenergetic beam of electrons and showed once again that it did not obey an exponential absorption law. He then modified the beam and made it heterogeneous by allowing it to pass through a thin sheet of platinum before striking an aluminum absorber. This resulted in an exponential absorption curve (Figure 8.20) similar to the one he had calculated previously:

The fact that β-rays, initially homogeneous, are absorbed according to an exponential law after passing through a small thickness of platinum has been confirmed, and it has been shown that this is not due to mere scattering of the rays, but to the fact that the beam is rendered heterogeneous in its passage through the platinum. (p. 325)

Conclusion

The controversy concerning the exponential absorption of electrons is a case of apparent, not real, discord. It was ultimately shown that both sets of experimental results were correct. The discord arose because one set of results had been misinterpreted: Schmidt and Meitner, Hahn, and von Baeyer believed that they were observing the absorption of mono-energetic electrons, when, in fact, they were observing the absorption of electrons with a continuous energy spectrum. Wilson, with some eviden-tial support from others, demonstrated clearly that monoenergetic elec-trons were not absorbed exponentially. He argued persuasively for the correctness of his results. He also demonstrated, first by calculation then by direct experiment, that exponential absorption actually showed that the electrons had a continuous spectrum of energies. Wilson and others had also shown that electrons lost energy when passing through matter, a critical assumption in his interpretation of the results. The evidence convinced the physics community that Wilson was correct. The issue was resolved by a critical examination and reinterpretation of the available experimental evidence.

9

The Liquid-Scintillator Neutrino Detector
Two Different Results from One Experiment

Beginning in the 1960s, physicists were faced with a discrepancy between the observed and theoretically calculated flux of neutrinos from the sun (the "solar neutrino problem").[1] Bahcall and Davis (1989)[2] later summarized the situation:

It is surprising to us, and perhaps more than a little disappointing, to realize that there has been very little qualitative change in either the observations or the standard theory since these papers[3] appeared, despite a dozen years of reexamination and continuous effort to improve details. (p. 508)

The reason for their disappointment is shown clearly in Figure 9.1, which depicts the experimental results and theoretical predictions as functions of time: Experiment and theory are unreconciled over the 16-year interval charted.

Various attempts to solve the problem by modifying the theory failed. Two alternative explanations were offered. The first was the idea of neutrino decay. If neutrinos have a finite lifetime, the solar neutrino deficit could be explained by the decay of significant numbers of neutrinos before they reach the earth.[4] This suggestion was rejected when neutrinos from the supernova SN 1987A were observed at the earth (Hirata et al., 1988):

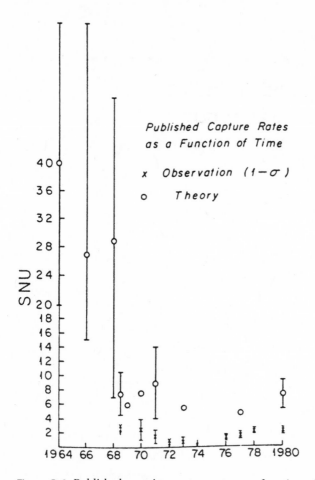

Figure 9.1. Published neutrino capture rates as a function of time. The circles are the theoretical predictions. The x's are the observed rates. From Bahcall and Davis (1989).

The idea that neutrino decay into some sterile form[5] might provide an explanation of the solar neutrino problem died in its most straightforward form along with the supernova SN 1987A, since the observation of (anti)neutrinos from that stellar explosion clearly requires survival times much longer than the Sun to Earth transit. (Anselmann et al., 1992, p. 395)

The second suggested alternative was neutrino oscillations—the idea that one type of neutrino can transform into another type.[6] For example, the electron neutrino might transform into a muon neutrino, and vice versa. This could explain the discrepancy between theory and experiment because the initial solar neutrino experiments were sensitive only to electron neutrinos: The latter would be undercounted if, during their travel from the sun to the earth, they transformed into undetectable muon or tau neutrinos.

There have been numerous experimental attempts to observe neutrino oscillations and at present, there is good experimental evidence for such oscillations.[7] The most recent measurement of the solar neutrino flux not only supports the hypothesis of neutrino oscillations, but also gives a result in agreement with the theoretical predictions. In this chapter, I discuss one of the early accelerator experiments that was designed to search for electron-neutrino oscillations. This was the Liquid-Scintillator Neutrino Detector (LSND) experiment. This experiment is of interest because different members of the experimental group initially produced discordant results by analyzing the same data in different ways—an almost unprecedented event in physics.

The LSND Experiment

The LSND experiment is an example of a neutrino-oscillation experiment that uses neutrinos produced by a high-energy particle accelerator. The experiment used pions produced by protons from the 800-MeV linear proton accelerator at Los Alamos National Laboratory. The pions then provided a source of muon antineutrinos by the process $\pi^+ \rightarrow \mu^+ + \nu_\mu$, followed by muon decay at rest, $\mu^+ \rightarrow e^+ + \nu_e + \bar{\nu}_\mu$.[8] The experimenters then searched for neutrino oscillations $\bar{\nu}_\mu \rightarrow \bar{\nu}_e$ by looking for the signature of the electron antineutrino. The signature was established by detection of both the positron and the neutron produced in the reaction $\bar{\nu}_e + p \rightarrow e^+ + n$, followed by 2.2-MeV γ rays produced by the reaction $n + p \rightarrow d + \gamma$.

It was, of course, extremely important that there be no possible source of electron antineutrinos present other than those produced by $\nu_\mu \leftrightarrow \nu_e$ oscillations. There was an obvious source of such electron antineutrinos resulting from the symmetrical decay chain starting with π^- mesons, which are also produced by the accelerator:

This background is suppressed by three factors in this experiment. First, π^+ production is about 8 times the π^- production in the beam stop. Second, 95% of π^-

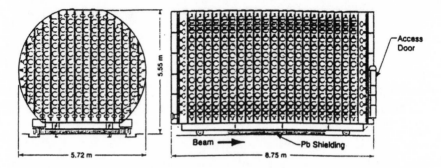

Figure 9.2. Schematic drawings of the LSND detector. From Athanassopoulos et al. (1997).

come to rest and are absorbed before decay in the beam stop. Third, 88% of μ^- from π^- DIF [decay in flight] are captured from atomic orbit, a process which does not give a $\bar{\nu}_e$. Thus, the relative yield, compared to the positive channel, is estimated to be $\sim(1/8) \times 0.05 \times 0.12 = 7.5 \times 10^{-4}$. A detailed Monte Carlo simulation gives a value of 7.8×10^{-4} for the flux ratio of $\bar{\nu}_e$ to $\bar{\nu}_\mu$. (Athanassopoulos et al., 1996a, p. 3082)

The detector, shown schematically in Figure 9.2, is a cylindrical tube 8.3 m long by 5.7 m in diameter, filled with 167 metric tons of liquid scintillator and viewed by 1220 phototubes. The phototubes detected both the Čerenkov radiation produced by relativistic positrons and the scintillation light. The tube was surrounded by an anticoincidence veto shield, also filled with liquid scintillator. The entire detector was shielded by 2 kg/cm^2 of overburden to reduce background due to cosmic rays. The timing and pulse heights of the photomultiplier pulses were used to reconstruct the electron or positron track (the detector did not distinguish between them). The relativistic positrons were detected by using the cone of Čerenkov light combined with the time distribution of the light from the phototubes, which is broader for nonrelativistic particles. The information allowed the experimenters to reconstruct the position and the time of the event. The apparatus also detected the delayed γ rays produced by the absorption of the neutron produced. The experimenters required that the positron energy be between 36 and 60 MeV, to reduce background due to the reaction $\nu_e + {}^{12}C \rightarrow e^- + X$ and to include all neutrinos with an energy up to the maximum allowed. In addition, they required that the event be more than 35 cm from the boundary of the detector. Once a positron had been identified, a search was made for an associated 2.2-MeV γ rays.

The physicists could determine Δr, the reconstructed distance between the γ ray and the positron; Δt, the relative time between detection of the γ ray and the positron; and Nγ, the number of phototubes triggered by the γ ray. They required $\Delta r < 2.5$ m, $\Delta t < 1$ ms (the absorption time was 186 μs), and Nγ (which is a measure of the γ-ray energy) be between 21 and 50. They then defined the function R(Δr, Δt, Nγ), which was approximately the ratio of the likelihood that the γ ray was correlated with the positron to the likelihood that it was an accidental coincidence. They found that "a γ ray with $R > 30$ has an efficiency of 23% for events with a recoil neutron and an accidental rate of 0.6% for events with no recoil neutron" (Athanassopoulos et al., 1995, p. 2651).

For $R > 30$ and a positron energy between 36 and 60 MeV, the group found nine beam-on events and 17 beam-off events. The beam status was not used in the trigger decision, but was recorded for each event. This allowed the experimenters to determine whether the event was associated with the beam and to study beam-unrelated backgrounds. The relative beam-on duty factor of 7.3% allowed 13 times more beam-off than beam-on data to be collected. This reduced the beam-off background to 17/13 or 1.3 events, giving a beam-on excess of 7.7 events. The background due to accidental γ rays with $R > 30$ was 0.79 ± 0.12, which gave a net excess of 6.9 events:

In conclusion, the LSND experiment observes 9 e$^+$ events within $36 < E_e < 60$ MeV, which satisfy strict criteria for a correlated low energy γ. The total estimated background from conventional processes is 2.1 ± 0.3 events, so the probability that the excess is a statistical fluctuation is $<10^{-3}$. If the excess obtained from a likelihood fit to the full e$^+$ sample arises from $\nu_\mu \leftrightarrow \nu_e$ oscillations, it corresponds to an oscillation probability of $(0.34^{+0.20}_{-0.18} \pm 0.07)\%$.[9] (p. 2653)

This was not, however, the conclusion reached by all of the members of the LSND group. Alfred Mann, one of the original collaborators on the experiment, withdrew from the group because he was concerned that the experimenters might be tuning their analysis procedure to produce a positive result.[10] He worried that the R criterion had been unconsciously shaped to find an effect and that it was unnecessarily complex. "My whole experience says that if you're going to find something new, it gen-

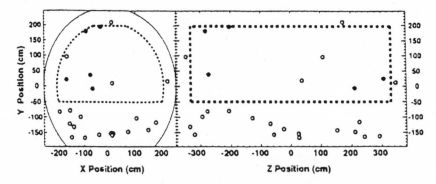

Figure 9.3. The 25 beam-on e-γ coincidences, before the application of the fiducial volume cut. Events within the fiducial region are denoted as solid circles, whereas those outside are represented as open circles. From Hill (1995).

erally rises up out of the data and pokes you in the eye" (Louis et al., 1997, p. 106).

James Hill, a graduate student working with Mann, presented an alternative analysis of the LSND data in a companion paper (Hill, 1995).[11] Rather than using R, he required $\Delta r < 2.4$ m, $\Delta t < 750$ μs ($\approx$ four capture times), and Nγ ≥ 26. These criteria were similar but not identical to those used by the rest of the LSND group in their analysis of the data. He noted that the background due to unrelated processes was extremely asymmetric, and was concentrated near the bottom of the detector:

The inhomogeneity of the background . . . and of the potential signal . . . requires confining the fiducial volume to a region of the detector that is not only more background free, but within which there are no strong gradients of event density. Since the backgrounds for both e± and coincident γ are inhomogeneous, and both enhanced at the bottom of the detector, the distribution of distance between the primaries and accidentally coincident γ rays will not be constant throughout the detector. This problem is addressed both by tightening the region analyzed and by the requirement that coincidences pass each of the separate criteria on the e± – γ relative time and distance, and γ-ray energy. (p. 2656)

Hill chose a more restricted fiducial volume for acceptable events, one that avoided the bottom of the detector (Figure 9.3). He found five events with an expected background of 6.2 ± 1.2 events, which "leaves no apparent signal for $v_\mu \leftrightarrow v_e$ oscillations" (p. 2656). He also checked that his conclusions and results were stable against reasonable variations in his selection criteria. He varied the fiducial volume, the Δr requirement, and the Δt requirement. His conclusion was always the same: There was no evidence of neutrino oscillations.

The discord within the LSND group concerning analysis procedure and data selection was not easy to resolve. The major difference between the two procedures was Hill's use of a more restrictive fiducial volume for accepted events, based on the distribution of background events. This seems to have been a matter of judgment. I note that blind analysis might have been usefully applied in this experiment. As we shall see in the next section, the group took Hill's criticism seriously and used his more restrictive volume in one of their analyses of subsequent data.

The Attempted Resolution

The LSND group continued taking data and reported new results based on all the data taken during 1993–1995 (Athanassopoulos et al., 1996a). (A more detailed analysis was presented in Athanassopoulos et al. [1996b].) They analyzed their data using several different selection cuts. These included the selection cuts used in their earlier analysis (Selection I) and a more relaxed set of cuts (Selection VI), which increased the signal efficiency by approximately 40%. The energy calibration, the energy resolution, and the particle identification scheme were checked using electrons from the decay of stopped muons. Cosmic-ray neutrons were used to check the properties of the 2.2-MeV γ rays. The group also addressed Hill's criticism concerning the detector background and presented an analysis of the data using the Selection VI cuts, but with a more restricted fiducial volume (Selection VIb). They remarked, "The second criterion defined as selection VIb, and motivated by [Hill's analysis] removes 55% of the acceptance" (Athanassopoulos et al., 1996b, p. 2699).

The results using all three sets of selection criteria are shown in Table 9.1. The group chose the Selection VI events to determine their final result. They concluded:

This paper reports the observation of 22 electron events in the $36 < E_e < 60$ MeV energy range that are correlated in time and space with a low-energy γ with $R > 30$, and the total estimated background from conventional processes is 4.6 $\pm$ 0.6 events. The probability that this excess is due to a statistical fluctuation is

Table 9.1.

The number of signal and background events in the $36 < E_e < 60$ MeV energy range

Selection[a]	Signal	Beam-Off[b]	ν Background	Excess	E/F[c]
I R $\geqslant$ 0	221	133.6 ± 3.1	53.5 ± 6.8	33.9 ± 16.6	130 ± 64
I R > 30	13	2.8 ± 0.4	1.5 ± 0.3	8.7 ± 3.6	146 ± 61
VI R $\geqslant$ 0	300	160.5 ± 3.4	76.2 ± 9.7	63.3 ± 20.1	171 ± 54
VI R > 30	22	2.5 ± 0.4	2.1 ± 0.4	17.4 ± 4.7	205 ± 64
VIb R $\geqslant$ 0	99	33.5 ± 1.5	34.3 ± 4.4	31.2 ± 11.0	187 ± 66
VIb R > 30	6	0.8 ± 0.2	0.9 ± 0.2	4.3 ± 2.5	110 ± 63

Source: Athanassopoulos et al. (1996a).

[a]VIb is a restrictive geometry test.

[b]The beam-off background has been scaled to the beam-on time.

[c]E/F is the excess number of events divided by the total efficiency.

4.1×10^{-8}. A fit to the full energy range $20 < E_e < 60$ MeV gives an oscillation probability of $(0.31 \pm 0.12 \pm 0.05)\%$. These results may be interpreted as evidence for $\nu_\mu \rightarrow \nu_e$ oscillations. (Athanassopoulos et al., 1996a, p. 3085).

There is still a noticeable positive effect for R > 0 and R > 30 using Selection VIb, the more restrictive fiducial volume. For R > 30, there are 4.3 ± 2.5 events. The probability that this is due to a fluctuation in the background of 1.7 ± 0.3 is 1.1%.

The LSND group provided a further check on their oscillation results by searching for a similar effect in $\nu_\mu \leftrightarrow \nu_e$ oscillations, using muon neutrinos obtained from pion decay in flight (DIF), a different experiment (Athanassopoulos et al., 1998a) with a more detailed analysis presented in Athanassopoulos and collaborators (1998b). Their original result had been obtained using ν_μ from muon decay at rest (DAR):

The analysis presented here uses a different component of the neutrino beam, a different detection process, and has different backgrounds and systematics from the previous DAR result, providing a consistency check on the existence of neutrino oscillations.[12] (Athanassopoulos et al., 1998a, p. 1774)

Once again, there had to be no possible source of electron neutrinos present other than those that might be produced by neutrino oscillation. Electron neutrinos from π^+ decay in flight were suppressed by the branching ratio of 1.23×10^{-4}, and those from μ^+ decay in flight were reduced by the longer muon lifetime and by the kinematics of the three-body decay of the muon ($\mu^+ \rightarrow e^+ + \nu_e + \overline{\nu}_\mu$).

Their previous experiment looked for the appearance of electron antineutrinos by identifying both the positron and the neutron produced. In the decay-in-flight experiment the electron neutrino was to be identified only by the electron produced in the reaction $\nu_e + C \rightarrow e^- + X$, in which the carbon nucleus was present in the liquid scintillator of the detector:

Candidate events for $\nu_\mu \leftrightarrow \nu_e$ oscillation from the DIF ν_μ flux consist of a single, isolated electron (from the [$\nu_e + C \rightarrow e^- + X$] reaction) in the energy range 60–200 MeV. The lower limit was chosen to be well above the end point of the Michel electron spectrum [from muon decay] (52.8 MeV) to avoid backgrounds induced by cosmic-ray muons and beam related ν_μ and $\overline{\nu}_\mu$ events. The upper limit of 200 MeV is the energy above which beam-off background rates increase, and the expected signal becomes much attenuated. *The analysis relies solely on electron PID [particle identification] in an energy region for which no control sample is available.* (Athanassopoulos et al. 1998a, p. 1775, emphasis added)

The electron identification scheme was crucial. The experimenters relied primarily on differences in the timing characteristics of the components of the light produced in the events: scintillation light, direct Čerenkov light, and rescattered Čerenkov light:

Each of the three light components has its own characteristic emission time distribution. The scintillation light has a small prompt peak plus a large tail which extends to hundreds of nanoseconds. The direct Čerenkov light is prompt and is measured with a resolution of approximately 1.5 ns. The scattered Čerenkov component has a time distribution between the direct Čerenkov light and the scintillation light, with a prompt peak and a tail that falls off more quickly than scintillation light. (Athanassopoulos et al., 1998b, p. 2495)

The experimenters used two different reconstruction algorithms to analyze the light and determine whether it fit the characteristics of that produced by a relativistic electron.

Background due to cosmic-ray muons was reduced by requiring fewer than four active hits in the veto shield (no charged particle present), a reconstructed distance of the event >35 cm from the boundary of the detector, and by cuts on the "space-time and multiplicity correlations

Table 9.2.

Comparison of results for the two analyses (A and B), their logical AND and OR[a]

Dataset	Beam On/Off	BUB[b]	BRB[c]	Excess	Efficiency (%)	Oscillation Probability ($\times 10^{-3}$)
A	23/114	8.0 ± 0.7	4.5 ± 0.9	10.5 ± 4.9	8.4	2.9 ± 1.4
B	25/92	6.4 ± 0.7	8.5 ± 1.7	10.1 ± 5.3	13.8	1.7 ± 0.9
AND	8/31	2.2 ± 0.3	3.1 ± 0.6	2.7 ± 2.9	5.5	1.1 ± 1.2
OR	40/175	12.3 ± 0.9	9.6 ± 1.9	18.1 ± 6.6	16.5	2.6 ± 1.0

Source: Athanassopoulos et al. (1998a).
[a]All errors are statistical.
[b]BUB is beam-unrelated background.
[c]BRB is beam-related background.

between the current event and its past/future neighboring events."[13] (Athanassopoulos et al., 1998a, p. 1775)

The candidate events were analyzed by two different analysis procedures, which had different reconstruction software and different selection criteria. The final samples generated by these two procedures did not have to be the same. The experimenters chose to use the logical "OR" events as their final sample. "This minimizes the sensitivity of the measurement to uncertainties in the efficiency calculation, is less sensitive to statistical fluctuations, and yields a larger efficiency" (Athanassopoulos et al., 1998a, p. 1776). Their final sample of events for each of the analysis procedures, along with the "AND" and "OR" samples is given in Table 9.2. They concluded:

We have described a search for $[v_e + C \rightarrow e^- + X]$ interactions for electron energies $60 < E_e < 200$ MeV. Two different analyses observe a number of beam-on events significantly above the expected number from the sum of conventional beam-related processes and cosmic-ray (beam-off) events. The probability that the 21.9 ± 2.1 estimated background events fluctuate into 40 observed events is 1.1×10^{-3}. The excess events are consistent with $v_\mu \leftrightarrow v_e$ oscillations with an oscillation probability of $(2.6 \pm 1.0 \pm 0.5) \times 10^{-3}$.[14] (Athanassopoulos et al., 1998a, p. 1777)

They further noted that "This $v_\mu \leftrightarrow v_e$ DIF oscillation search has completely different backgrounds and systematic error from the $v_\mu \leftrightarrow v_e$ DAR oscillation search and provides additional evidence that both effects are due to neutrino oscillations" (p. 1777).

Although each of the LSND results on neutrino oscillations appears persuasive and the two different results are consistent and mutually supportive, the fact remains that no other similar experiment has obtained a positive result.

The most precise results are from the Karmen experiment. The Karmen detector was a liquid-scintillator detector quite similar to LSND. The group searched for both $\bar{\nu}_\mu \leftrightarrow \bar{\nu}_e$ and $\nu_\mu \leftrightarrow \nu_e$ oscillations. In the former case, they looked for the same signature as had the LSND group: a spatially correlated, delayed coincidence between a positron produced in the reaction $\bar{\nu}_e + p \rightarrow e^+ + n$, and the γ rays produced when the neutron was absorbed in the detector. They found 124 candidate events, with a background of 96.7, giving an excess of 27.3 ± 11.4 events. With more stringent selection criteria, the excess signal was 7.8 ± 6.3 events, which they regarded as insignificant beam excess.

The signature chosen for $\nu_\mu \leftrightarrow \nu_e$ oscillations was the electron produced in the reaction $\nu_e + {}^{12}C \rightarrow {}^{12}N + e^-$ (the same reaction used by the LSND group in their DIF experiment), but with the additional requirement of a detected positron from the decay ${}^{12}N \rightarrow {}^{12}C + e^+ + \nu_e$. This was a more stringent set of criteria than that used by LSND. During an experimental run from July 1992 to December 1997, they found only two candidate events, with a calculated background of 2.26 ± 0.3. They concluded, for both searches, that "No evidence for oscillations could be found with Karmen" (Zeitnitz et al., 1998, p. 169).

According to a recent survey (Peltoniemi Web site):

All experiments, except LSND, are consistent with no oscillation. The results of LSND can be interpreted as a signal of oscillation of muon neutrinos to electron neutrinos. Most of the parameter range $[\sin^2 2\theta, \Delta m^2]$ explaining the LSND results are in disagreement with other experiments, particularly Karmen. However, there still seems to be a small area allowed by all experiments.

Until recently, there was convincing evidence only for $\nu_\mu \leftrightarrow \nu_\tau$ oscillations. This came primarily from experiments on atmospheric neutrinos.[15] In the summer of 2001, striking evidence was provided for electron-neutrino oscillations. The Sudbury Neutrino Observatory (SNO) reported that the flux of solar neutrinos from the decay of ${}^8B$, as measured by the charged current reaction $\nu_e + d \rightarrow p + p + e^-$ (a reaction sensitive only to electron neutrinos) was $\varphi^{CC}(\nu_e) = 1.75 \pm 0.07$ (statistical) $^{+0.12}_{-0.11}$ (systematic) ± 0.05 (theoretical) $\times 10^6$ cm^2 s^{-1}. The value of the ^{8}B neutrino flux measured by the Super-Kamiokonde Collaboration for the elastic scattering reaction $\nu_x + e^- \rightarrow \nu_x + e^-$ (which is sensitive to all three types of neutrinos) was $\varphi^{ES}(\nu_e) = 2.23 \pm 0.03$ (statistical) $^{+0.08}_{-0.07}$ (systematic) $\times 10^6$ cm^2 s^{-1}. The difference between the two measured fluxes indicates the presence of other types of neutrinos in the solar flux observed at the earth, and is strong evidence for electron-neutrino oscillations. "Comparison of $\varphi^{CC}(\nu_e)$ to the Super-Kamiokonde Collaboration's precision value of $\varphi^{ES}(\nu_e)$ yields a 3.3 σ [standard deviation] difference, providing evidence that there is a non-electron flavor active neutrino component in the solar flux"[16] (SNO Web site).

In addition, the SNO Collaboration determined that the total flux of active ^{8}B neutrinos was $5.44 \pm 0.99 \times 10^6$ cm^2 s^{-1}, in "excellent agreement" with the predictions of the Standard Solar Model of 5.05×10^6 cm^2 s^{-1}. Thus, the solar-neutrino problem seems to have been solved: The deficit was due to electron-neutrino oscillations.

Unfortunately, the SNO result does not cast any light on whether the positive LSND result on electron-neutrino oscillations is correct. As noted above, the theoretical analysis of the experimental results still allows a consistent interpretation of all the results, although the allowed region (in $\sin^2 2\theta$ and Δm) for the LSND result is shrinking. The question of the correctness of the LSND result must await the results of the BooNe experiment at Fermilab. "BooNE is motivated predominantly by the evidence for neutrino oscillations as claimed by the LSND experiment. The goals of BooNE are: 1) confirm (or refute) the LSND observation with much better statistical precision (thousands of events compared to tens of events in LSND)" (BooNE Web site).

Discussion

In this chapter, we have discussed an episode in which there were two different sets of discordant results. The first was within the LSND group. The second was the disagreement between the LSND result and the results of other accelerator experiments, most notably Karmen. Although the issues are still unresolved, I believe that this episode is interesting in itself and also merits discussion because the methods already used to try to resolve the issues are of interest.

The disagreement between the majority of the LSND group and Hill concerned data analysis and selection procedures. The LSND group responded to Hill's criticism by using his fiducial volume in one of their analyses of their later data. They argued that their result was robust against such changes in their selection criteria (see Table 9.1). They also performed a second experiment in which neutrinos from muon decays in flight rather than from muon decays at rest were used. The consistency of their results, despite the difference in sources of background and systematic errors, also argued for the robustness and correctness of their result.

Nevertheless, the failure of other experiments to reproduce the LSND result leaves the issue unresolved.[17] LSND claims only that the probability that their result is due to a fluctuation of background varies between .1% and 1%, depending on the result and the analysis procedure used. This is unlikely, but not impossible. The greatly improved statistical accuracy of the BooNE experiment should resolve the issue.

10

Atomic Parity Violation, SLAC E122, and the Weinberg-Salam Theory

The final episode I discuss involves discordant results from experiments that tested the Weinberg-Salam (W-S) unified theory of electroweak interactions. This is the most complex of the episodes considered because it involves two different (but intertwined) instances of discordant experimental results. First, there was a disagreement between different experiments that measured atomic parity violation. Some of these experiments seemed to refute the W-S theory, whereas others supported it. In addition, the question of whether these experimental results confirmed the W-S theory was dependent on atomic physics calculations, which were themselves uncertain. The second discord was between the early negative results from atomic parity–violation experiments and the supportive result for the W-S theory produced by the SLAC E122 experiment, which measured an asymmetry in the scattering of polarized electrons from deuterium. Thus we have discordant results produced in experiments that measured the same, or similar, quantities, by similar techniques. We also have discordant results between experiments that measured very different quantities by very different experimental methods, in which the discord centered on whether the experiments supported or refuted the same theory.

The early history of the episode may be summarized as follows. In 1957, it had been experimentally demonstrated that parity (i.e., left-right symmetry) was violated in the weak interactions.[1] This feature of weak interactions had been incorporated into the Weinberg-Salam unified theory of electroweak interactions. The theory predicted that one would see weak neutral-current effects in the interactions of electrons with hadrons (strongly interacting particles). The effect would be quite small when compared with the dominant electromagnetic interaction, but could be distinguished from it by the violation of parity conservation. A demonstration of a parity-violating effect and measurement of its magnitude would test the W-S theory. One such predicted effect was the rotation of the plane of polarization of polarized light when it passed through bismuth vapor.

In 1976 and 1977, experimental groups at Oxford University and the University of Washington reported results from atomic parity–violation experiments that disagreed with the predictions of the W-S theory. At the time the theory had other experimental support, but was not universally accepted. In 1978 and 1979, a group at the Stanford Linear Accelerator Center (the SLAC E122 experiment) reported results on the scattering of polarized electrons from deuterium, which confirmed the W-S theory. By 1979, the W-S theory was regarded by the high-energy physics community as established, although as Andrew Pickering (1984a) stated, "there had been no *intrinsic* change in the status of the Washington-Oxford experiments" (p. 301).[2]

In Pickering's (1984a) view:

particle physicists *chose* to accept the results of the SLAC experiment, *chose* to interpret them in terms of the standard model (rather than some alternative which might reconcile them with the atomic physics results) and therefore *chose* to regard the Washington-Oxford experiments as somehow defective in performance or interpretation. (p. 301)

The implication seems to be that these choices were made so that the experimental evidence would be consistent with accepted theory, and that there were no good, independent reasons for the decision. In other words, the disagreement was not resolved on epistemological or methodological grounds, but rather by loyalty to existing community commitments.

My view is quite different. I regard the two experimental results as having different evidential weights. The initial Washington-Oxford results used new and untested experimental apparatus and had large systematic uncertainties (as large as the predicted effects). In addition, their initial results, reported in 1976 and 1977, were internally inconsistent and by 1979 there were other atomic parity–violation results that confirmed the W-S theory. Thus by the end of 1979, the overall situation with respect to the atomic parity results was quite uncertain. In contrast, the SLAC experiment, although also using new techniques, had been very carefully checked and had far more evidential weight.

Faced with this situation, the physics community chose to accept the SLAC results, which supported the W-S theory, and to await further developments on the uncertain atomic parity–violation results. The experiment-theory disagreement and the discord between the two sets of experimental results on atomic parity violation were later resolved, as was the disagreement between the early atomic parity–violation results and those of SLAC E122. Both disagreements were resolved by reasoned argument based on experimental evidence and epistemological and methodological criteria. I argue for this by a detailed examination of the history of this episode.

Early Atomic Parity–Violation Experiments

The first experimental tests of the W-S theory were performed by groups at Oxford and Washington. They looked for a parity-violating rotation of the plane of polarization of light when it passed through bismuth vapor. They both used bismuth vapor but used wavelengths corresponding to different transitions in bismuth: $\lambda = 648$ nm (Oxford) and $\lambda = 876$ nm (Washington). They published a joint preliminary report noting that "we feel that there is sufficient interest to justify an interim report" (Baird et al., 1976, p. 528). They reported values for R, the parity-violating parameter, of $R = (-8 \pm 3) \times 10^{-8}$ (Washington) and $R = (+10 \pm 8) \times 10^{-8}$ (Oxford). "We conclude from the two experiments that the optical rotation, if it exists, is smaller than the values -3×10^{-7} and -4×10^{-7} predicted by the W-S model plus the atomic central field approximation" (p. 529).[3]

The experimental results were quite uncertain; they included systematic uncertainties of the order of $\pm 10 \times 10^{-8}$ that were not fully understood. The systematic experimental uncertainties were of the same order of magnitude as the expected effect. These were also novel experiments, using new and previously untried techniques, which also tended to make the experimental results uncertain.

In September 1977, both the Washington and Oxford groups published more detailed accounts of their experiments with somewhat revised results (Baird et al., 1977; Lewis et al., 1977). Both groups again reported results in substantial disagreement with the predictions of the W-S theory, although the Washington group stated that "more complete calculations that include many-particle effects are clearly desirable" (Lewis et al., 1977, p. 795). The Washington group reported a value of R = (−0.7 ± 3.2) × 10^{-8}, which was in disagreement with the theoretical prediction of approximately −2.5 × 10^{-7}. This value was also inconsistent with their earlier result of (−8 ± 3) × 10^{-8}.[4] This inconsistency was not discussed by the experimenters in the published paper, but it was discussed by others within the atomic-physics community. It lessened the credibility of the result.[5] The Oxford result was R = (+2.7 ± 4.7) × 10^{-8}, again in disagreement with the W-S prediction of approximately −2.5 × 10^{-7}. They noted, however, that there was a systematic effect in their apparatus. They found a change in the rotation angle φ (due to slight misalignment of the polarizers, optical rotation in the windows, etc.) of order 2 × 10^{-7} radians:

Unfortunately, it varies with time over a period of minutes, and depends sensitively on the setting of the laser and the optical path through the polarizer. While we believe we understand this effect in terms of imperfections in the polarizers combined with changes in laser beam intensity distribution, we have been unable to reduce it significantly. (Baird et al., 1977, p. 800)

A systematic effect of the same size as that of the theoretically predicted effect cast doubt on the result and on the comparison between experiment and theory.

The theoretical calculations of the expected effect were also uncertain. The problem is that for an atom with few electrons, where the electron wavefunctions could be calculated quite reliably, the predicted effect is small. For a multi-electron atom such as bismuth, in which the predicted effect is much larger, the wavefunctions can be calculated only approximately and with a fair amount of uncertainty. There were, at the time, four different calculations of the expected effect that agreed with one another to within ±25% (Table 10.1), so that the largest and smallest calculated values of R differ by almost a factor of two. The experimenters thought this rough agreement encouraging, although they did not know whether inclusion of the many-body effects (which had been

Table 10.1.
Calculated parity–violation effect in bismuth

Method	$R(10^{-7})$	Reference
Hartree-Fock	−2.3	Brimicombe et al. (1976)
Hartree-Fock	−3.5	Henley and Wilets (1976)
Semiempirical	−1.7	Novikov et al. (1976)
Multiconfiguration	−2.4	Grant (1976, pers. comm.)

Source: Lewis et al. (1977).

neglected in the calculation) would resolve the discrepancy between theory and experiment.

How were these results viewed at the time by the physics community? In the same issue of *Nature* in which the original joint Oxford-Washington paper was published, Frank Close, a particle theorist, summarized the situation:

Is parity violated in atomic physics? According to experiments being performed independently at Oxford and the University of Washington the answer may well be no. . . . This is a very interesting result in light of last month's report . . . claiming that parity is violated in high energy "neutral current" interactions between neutrinos and matter. (Close 1976, p. 505)

The experiment that Close referred to had concluded:

Measurements of R^ν and $R^{\bar{\nu}}$, the ratios of neutral current to charged current rates for ν and $\bar{\nu}$ [neutrino and antineutrino] cross sections, yield neutral current rates for ν and $\bar{\nu}$ that are consistent with a pure V-A interaction but 3 standard deviations from pure V or pure A, indicating the presence of parity nonconservation in the weak neutral current. (Benvenuti et al., 1976, p. 1030)

Close (1976) noted that the atomic-physics results appeared to be inconsistent with the predictions of the W-S model supplemented by atomic-physics calculations. He also remarked that "At present the discrepancy can conceivably be the combined effect of systematic effects in atomic physics calculations and systematic uncertainties in the experiments" (pp. 505–6). Close discussed the possibility that neutral-current effects might violate parity in neutrino interactions and conserve parity in electron interactions. He also discussed an alternative that had an unexpected (on the basis of accepted theory) energy dependence, so that the high-energy neutrino experiments showed parity nonconservation whereas the low energy atomic physics experiments would not. "Whether such a possibility could be incorporated into the unification ideas is not clear. It also isn't clear, yet, if we have to worry. However, the clear blue sky of summer now has a cloud in it. We wait to see if it heralds a storm" (p. 506).

The uncertainty caused by these atomic parity–violation results is shown in a summary of the Symposium on Lepton and Photon Interactions at High Energies, held in Hamburg August 25–31, 1977, given by David Miller. Miller (1977) noted that Sandars had reported that neither his group at Oxford nor the Washington group had seen any parity-violating effects and that "they have spent a great deal of time checking both their experimental sensitivity and the theory in order to be sure" (p. 288). Miller continued:

S. Weinberg and others discussed the meaning of these results. It seems that the SU(2) × U(1) is to the weak interaction what the naive quark-parton model has been to QCD, a first approximation which has fitted a surprisingly large amount of data. Now it will be necessary to enlarge the model to accommodate the new quarks and leptons, the absence of atomic neutral currents, and perhaps also whatever it is that is causing trimuon events. (p. 288)

I believe, however, that the uncertainty in these experimental results only made the disagreement with the W-S theory a matter of concern— not a full-blown crisis. In any event, the monopoly of Washington and Oxford was soon broken.

The evidential situation changed in 1978, when Barkov and Zolotorev (1978a,b, 1979a), two Soviet scientists from Novosibirsk, reported measurements on the same transition in bismuth as the Oxford group had studied. Their results agreed with the predictions of the W-S model. They gave a value for $\psi_{exp}/\psi_{W-S} = (+1.4 \pm 0.3)$ k, where ψ is the angle of rotation of the plane of polarization caused by the bismuth vapor. "The factor k was introduced because of inexact knowledge of the bismuth vapor, and also because of some uncertainty in the estimate, the factor lies in the interval from 0.5 to 1.5" (Barkov and Zolotorev, 1978a, p.360). They concluded that their result did not contradict the predictions of the W-S model. Note that agreement with theoretical prediction depended (and still does depend) on which method of calculation is chosen. A somewhat later paper changed the result to $\psi_{exp}/\psi_{W-S} = 1.1 \pm 0.3$ (Barkov and Zolotorev, 1978b).

Subsequent papers reported more extensive data and found a value for $R_{exp}/R_{theor} = 1.07 \pm 0.14$ (Barkov and Zolotorev, 1979a,b, 1980). They also reported that the latest unpublished results from the Washington and Oxford groups, which had been communicated to them privately, showed parity violation, although "the results of their new experiments have not reached good reproducibility" (Barkov and Zolotorev, 1979a, p. 312).

In September 1979, an international workshop devoted to neutral-current interactions in atoms was held in Cargese, France. This workshop was attended by representatives of virtually all of the groups actively working in the field, including Oxford, Washington, and Novosibirsk. At that workshop, the Novosibirsk group presented a very detailed account of their experiment (Barkov and Zolotorev 1979b). Bouchiat (1980) remarked in his workshop summary paper, "Professor Barkov, in his talk, gave a very detailed account of the Novosibirsk experiment and answered many questions concerning possible systematic errors" (p. 364). There was also communication between the Soviet and Oxford groups. The Soviets reported that they had been able to uniquely identify the hyperfine structure of the 6477-Å [648 nm] line of atomic bismuth and that "the results of these measurements agree also with the results in Oxford (P. Sandars, pers. comm.)" (Barkov and Zolotorev 1978a, p. 359).

In early 1979, a Berkeley group reported an atomic-physics result for thallium that agreed with the predictions of the W-S model (Conti et al., 1979). They investigated the polarization of light passing through thallium vapor and found a circular dichroism $\delta = (+5.2 \pm 2.4) \times 10^{-3}$ and compared this with the theoretical prediction of $(+2.3 \pm 0.9) \times 10^{-3}$. Although these were not definitive results—they were only two standard deviations from zero—they did agree with the model in both sign and magnitude.

In mid-1979, the attempts to confirm W-S theory through atomic physics experiment were inconclusive. The Oxford and Washington groups had originally reported a discrepancy between their experimental results and the theory, but their more recent results, although preliminary, showed the presence of the predicted parity nonconserving effects. In addition, the Soviet and Berkeley results agreed with the model. Dydak (1979) summarized the situation in a talk at a 1979 conference:

It is difficult to choose between the conflicting results in order to determine the *eq* [electron-quark] coupling constants. Tentatively, we go along with the positive results from Novosibirsk and Berkeley groups and hope that future development will justify this step (it cannot be justified at present, on clear-cut experimental grounds).[6] (p. 35)

Bouchiat's (1980) summary paper at the Cargese Workshop was more positive. After reviewing the Novosibirsk experiment as well as the conflict between the earlier and later Washington and Oxford results he remarked, "*As a conclusion on this Bismuth session, one can say that parity violation has been observed roughly with the magnitude predicted by the Weinberg-Salam theory*" (p. 365, emphasis in original). But even this statement does not assert that the results agree with the predictions of the theory.[7]

The SLAC E122 Experiment

The evidential situation was made even more complex when a group at SLAC reported a result from the SLAC E122 experiment on the scattering of polarized electrons from deuterium that agreed with the W-S model (Prescott et al., 1978, 1979). They not only found the predicted scattering asymmetry but also obtained a value for $\sin^2\theta_W = 0.20 \pm 0.03$ (1978) and 0.224 ± 0.020 (1979) in agreement with other measurements made at the time ($\sin^2\theta_W$ is an important parameter in the W-S theory). "We conclude that within experimental error our results are consistent with the W-S model, and furthermore our best value of $\sin^2\theta_W$ is in good agreement with the weighted average for the parameter obtained from neutrino experiments" (Prescott et al., 1979, p. 528).

Let us examine the arguments presented by the SLAC group in favor of the validity and reliability of their measurement. I agree with Pickering (1984a) that "in its own way E122 was just as innovatory as the Washington-Oxford experiments and its findings were, in principle, just as open to challenge" (p. 301). For this reason, the SLAC group presented a very detailed analysis of their experimental apparatus and result and performed many checks on their experiment.

The experiment depended, in large part, on a new high intensity source of longitudinally polarized electrons. The polarization of the electron beam could be varied by changing the voltage on a Pockels cell. "This reversal was done randomly on a pulse to pulse basis. The rapid reversals minimized the effects of drifts in the experiment, and the randomization avoided changing the helicity synchronously with periodic changes in experimental parameters" (Prescott et al. 1978, p. 348). It had been demonstrated in an earlier experiment that polarized electrons could be accelerated with negligible depolarization. In addition, both the sign and magnitude of the beam polarization were measured periodically by observing the known asymmetry in elastic electron-electron scattering from a magnetized iron foil.

The experimenters also checked whether the apparatus produced spurious asymmetries. They measured the scattering using the unpolarized beam from the ordinary SLAC electron gun, for which the asymmetry should be zero. They assigned polarizations to the beam using the same random-number generator that determined the sign of the voltage on the Pockels cell. They obtained a value for $A_{exp}/P_e = (-2.5 \pm 2.2) \times 10^{-5}$, where A_{exp} is the experimental asymmetry and P_e is the beam polarization for the polarized source ($P_e = 0.37$). This value is consistent with zero and demonstrated that the apparatus could measure asymmetries of the order of 10^{-5}.

They also varied the polarization of the beam by changing the angle of a calcite prism within the device, thereby changing the polarization of the light striking the Pockels cell. They predicted $A_{exp} = |P_e| A \cos(2\varphi_p)$, where φ_p was the prism angle. The results are shown in Figure 10.1. Not only do the data fit the predicted curve, but the fact that the results at 45° are consistent with zero indicates that other sources of error in A_{exp} are small. The graph shows the results for two different detectors, a nitrogen-filled Čerenkov counter and a lead glass shower counter. The consistency of the results confers credibility on the measurements: "Although these two separate counters are not statistically independent, they were analyzed with independent electronics and respond quite differently to potential backgrounds. The consistency between these counters serves as a check that such backgrounds are small" (Prescott et al. 1978, p. 350).

Because of the g-2 precession of the spin as the electrons passes through the beam transport magnets, the electron-beam helicity also depended on the beam energy E_0. The expected distribution and the experimental data for $A_{exp}/|P_e|Q^2$ are shown in Figure 10.2 (Q^2 is the square of the momentum transfer). It was noted that the data follow the

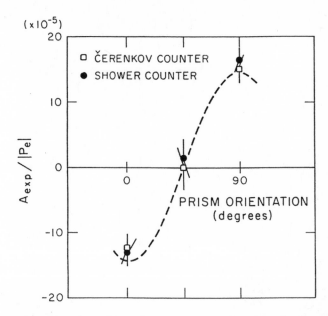

Figure 10.1. Experimental asymmetry as a function of prism angle for both the Čerenkov counter and the shower counter. The dashed line is the predicted behavior. From Prescott et al. (1978).

g-2 modulation of the helicity; the nearly zero value at 17.8 GeV demonstrated that any transverse spin effects were small.

A potentially serious source of error came from the possibility of small, systematic differences in the beam parameters for the two helicities. Small changes in beam position, angle, current, or energy could influence the measured yield, and if correlated with reversals of beam helicity could cause spurious asymmetries resembling parity-violating effects. These quantities were carefully monitored and a feedback system used to stabilize them:

Using the measured pulse to pulse beam information together with the measured sensitivities of the yield to each of the beam parameters, we made corrections to the asymmetries for helicity dependent differences in beam parameters. For these corrections, we have assigned a systematic error equal to the correction itself. The most significant imbalance was less than one part per million in E_0 [the beam energy] which contributed -0.26×10^{-5} to A/Q^2." (Prescott et al., 1978, p. 351) Compare this value with their final result of $A/Q^2 = (-9.5 \pm 1.6)$

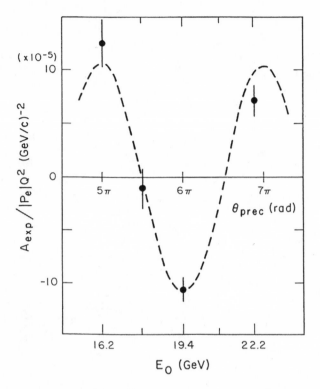

Figure 10.2. Experimental asymmetry as a function of beam energy. The expected behavior is the dashed line. From Prescott et al. (1978).

$\times 10^{-5}$ GeV/c^2. This was regarded by the physics community as a reliable and convincing result.[8]

Hybrid models, which might have reconciled the discordant atomic parity results and the results of SLAC E122, were both considered and tested by the E122 group. In their first paper (Prescott et al., 1978), they pointed out that the hybrid model was consistent with their data only for values of $\sin^2\theta_W < 0.1$, which was inconsistent with the measured value of approximately 0.23. In their second paper (Prescott et al., 1979) they plotted their data as a function of $y = (E_0 - E')/E_0$, where E' is the energy of the scattered electron. Both the W-S theory and the hybrid model made definite predictions for this graph. The results are shown in Figure 10.3: The superiority of the W-S model is obvious. For the W-S theory, they obtained a value of $\sin^2\theta_W = 0.224 \pm 0.020$ with a χ^2 probability of

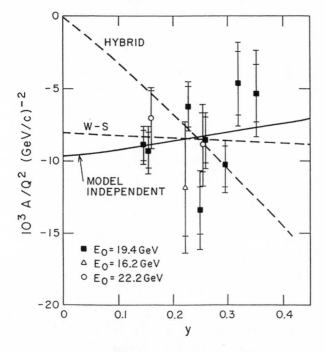

Figure 10.3. Asymmetries measured at three different energies plotted as a function of $y = (E_0 - E')/E_0$. The predictions of the hybrid model, the W-S theory, and a model-independent calculation are shown. "The Weinberg-Salam model is an acceptable fit to the data; the hybrid model appears to be ruled out." From Prescott et al. (1978).

40%. The hybrid model gave a value of 0.015 and a χ^2 probability of 6×10^{-4}, which, the experimenters noted, "appears to rule out this model."

The physics community chose to accept the results from the carefully conducted and checked SLAC experiment that confirmed the W-S theory, and to await further developments in the atomic parity–violating experiments, which, as we have seen, were inconclusive. This view is supported by Bouchiat (1980). After hearing a detailed account of the SLAC experiment, he stated:

[In] our opinion, this experiment gave the first truly convincing evidence for parity violation in neutral current processes. . . . I would like to say that I have been very much impressed by the care with which systematic errors have been treated in the experiment. It is certainly an example to be followed by all people working in this very difficult field. (pp. 358, 359–60)

The decision to accept the SLAC E122 result was based on evidential weight, determined by epistemological criteria.

Later Atomic Parity–Violation Experiments and the Resolution of the Discord

I discussed earlier the uncertainty in the 1977 Washington and Oxford results caused by systematic effects. This uncertainty is emphasized by examining the reports of Bogdanov and collaborators (1980a,b) of measurements on the same transition in bismuth that the Oxford group had used.[9] Their measurement was also in disagreement with the predictions of the W-S theory. They reported an optical rotation due to the parity-nonconservation interaction of $\varphi_{PNC} = (-0.22 \pm 1.0) \times 10^{-8}$ rad, in disagreement with the theoretical prediction of 10^{-7} rad. They discussed two sources of systematic errors that could give rise to effects of the same size as those expected from parity nonconservation: variation in laser intensity due to scanning the laser frequency and interference between the main laser beam and scattered light. Bogdanov and collaborators also discussed the measures taken to reduce the errors due to these effects. They remarked that the spread in the individual series of measurements substantially exceeded the error in their quoted result and attributed it to time-dependent instrumental errors. Once again there were systematic errors in this type of experiment that were approximately the same size as the effects predicted by the W-S theory.

The Washington group (Hollister et al., 1981) subsequently emphasized the discord between the various experimental results: "Our experiment and the bismuth optical-rotation experiments by three other groups [Oxford, Moscow, and Novosibirsk] *have yielded results with significant mutual discrepancies far larger than the quoted errors*" (p. 643, emphasis added). They also pointed out that their earlier measurements were not mutually consistent, empasizing the uncertainty in the results. The moral of the story is clear: These were extremely difficult experiments, beset with systematic errors of approximately the same size as the predicted effects.

Let us briefly examine the subsequent history of the bismuth experiments, along with a brief treatment of the other atomic physics parity–violation experiments that have relevance for the W-S theory.

Bourchiat's (1980) summary of the situation is shown in Table 10.2. Bouchiat remarked that whereas the Novosibirsk result had been pub-

Table 10.2.
Bismuth optical activity ($\lambda = 648$ nm)

Experiment	R ($\times 10^8$)	Theory
Novosibirsk	-20.6 ± 3.2	Novosibirsk -19
Oxford I	-10.3 ± 1.8	Oxford -14
Oxford II	-11.2 ± 4.1	
Old Oxford	2.7 ± 4.7	
($\lambda = 876$ nm)		
Washington	-10.0 ± 2	Novosibirsk -14
		Oxford -12
Old Washington	-0.7 ± 3.2	

Source: Bouchiat (1980).

lished, both the Washington and Oxford results were in the nature of progress reports on recent trends in their experiments, not definite results. He also noted that there was no explanation of the large difference between the old and new Washington and Oxford results, and that there was a factor of two discrepancy between the Novosibirsk and Oxford results at $\lambda = 648$ nm. The difference in both theoretical approach and the numerical value of the calculation between the two groups was also mentioned.

In 1981, the Washington group published another measurement of the optical rotation in bismuth at $\lambda = 876$ nm (Hollister et al., 1981). The theoretical calculations they used to compare with their data ranged from $R = -8 \times 10^{-8}$ to -17×10^{-8}. Their value of $R = (-10.4 \pm 1.7) \times 10^{-8}$ agreed in "sign and approximate magnitude with recent calculations of the effect in bismuth based on the Weinberg-Salam theory." They pointed out that since making their earliest measurements, they had "added a new laser, improved the optics, and included far more extensive experimental checks" (p. 643). They excluded the first three measurements from their average because they were made without the new systematic checks and controls, which were to be discussed in detail in a forthcoming paper.

That discussion appeared in an extensive review of atomic parity–violation experiments by Fortson and Lewis (1984), two members of the Washington group. They reported experimental controls on both the polarizer angle and the laser frequency. They also used alternate cycles, in which their bismuth oven was turned off, to avoid a spurious effect that could mimic the predicted parity-nonconserving effect. They also examined their data for any correlations between the measured values of the parity-nonconserving parameter and any other experimental variables. This procedure set limits on known sources of systematic error and initially helped to uncover some errors and eliminate them. These errors included those due to wavelength-dependent effects and to beam movement, errors common to all atomic parity experiments.

The Novosibirsk group's results did not change very much from the value cited above. Their last published measurement (Barkov and Zolotorev, 1980) gave $R = (-20.2 \pm 2.7) \times 10^{-8}$, which was approximately twice the value obtained by the Oxford group for the $\lambda = 648$ nm transition. The Moscow group (Bogdanov et al., 1980a) had originally reported a value for R in disagreement with the theoretical predictions and the experimental results of both the Oxford and Novosibirsk groups. Their value was $R_{exp}/R_{th} = -0.02 \pm 0.1$. A second publication (Bogdanov et al., 1980b) reported a value of $R = (-2.4 \pm 1.3) \times 10^{-8}$, still in disagreement with both theory and the other experimental measurements. They noted, however, that the errors within an individual series of measurements exceeded the standard deviation in some cases, indicating that there were additional systematic errors present that varied comparatively slowly with time. The Moscow group continued their investigation of the sources and magnitudes of systematic errors (Birich et al., 1984). They took steps to minimize these effects and to measure any residual effects, and noted that their earlier results had not included all of these controls and corrections. Their final value was $R = (-7.8 \pm 1.8) \times 10^{-8}$ and they concluded that "It is clear that our latest results and the results of the Oxford group [Oxford was reporting a value of approximately $(-9 \pm 2) \times 10^{-8}$ at this time] are in sufficient agreement with one another and with the results of the most detailed calculations" (p. 448).

The Oxford group continued their measurements on the $\lambda = 648$ nm line in bismuth through 1987. They had presented intermediate reports at conferences in 1982 and 1984 consistent with the W-S theory. Their 1987 result of $R = (-9.3 \pm 1.4) \times 10^{-8}$ (Taylor et al., 1987) is consistent with the standard model (the uncertainty is now primarily in the theoretical calculations) and with the measurements of Birich and collaborators (1984), but inconsistent with those of Barkov and Zolotorev (1979b). During the 1980s, the group devoted considerable effort to searching for systematic effects and trying to eliminate or correct for them. The difficulties of this type of experiment are severe: They noted that their method depends on changing the wavelength of the laser and that wavelength-dependent angles (WDA), comparable to the expected parity-nonconserving optical-rotation angle, are seen in their apparatus, even in the absence of bismuth.[10] This WDA varied with time in an apparently random way and affected the group's ability to make measurements and carry out diagnostic tests. Because of its random nature, they did not expect the WDA to give rise to systematic error in the bismuth measurements, but its presence does indicate the possibility of angle effects of similar size to the expected effect and hence the need for caution. Taylor and collaborators (1987) list experimental checks for angle sensitivity, angle lock, polarizer reversal, Faraday contamination, pickup, cross-modulation between laser and magnetic field, transverse magnetic field effects, and oven reversal. As one can see, making a valid measurement demands considerable care. They noted that their present result disagreed with their earlier published value of $R = (+2.75 \pm 4.7) \times 10^{-8}$. Because their new result involved an improved apparatus, considerably more data, and numerous checks against possible systematic error, they preferred their latest result. They concluded that their earlier result was in error, but conceded that they did not have any explanation for the source of the error. The rebuilding of the apparatus precluded testing many of the likely explanations.

The situation today is virtually the same as when Bouchiat and Pottier (1984) presented their summary (Table 10.3). The bismuth results are in approximate agreement with the W-S theory, although the discrepancy with the Novosibirsk measurement remains problematic. There are recent reports of a new Novosibirsk experiment whose results agree with those of Oxford (S. Blundell, pers. comm.).

Atomic parity violation has also been observed in elements other than bismuth. I mentioned earlier an experiment on thallium (Conti et al., 1979), which had given a result in approximate agreement with the W-S theory. The early bismuth experiments had been in part evaluated in the context of this experiment. The experiment had measured the circular dichroism, δ, and had found $\delta = (+5.2 \pm 2.4) \times 10^{-3}$ to be in agreement with the theoretical value $\delta_{th} = (+2.3 \pm 0.9) \times 10^{-3}$. Bucksbaum and collaborators continued this series of experiments through the 1980s, using the same basic method, although they made improvements

Table 10.3.
Bismuth optical activity ($\lambda = 648$ nm)

Experiment	R ($\times 10^8$)	Theory	Reference
Novosibirsk (1979)	-20.2 ± 2.7	-13	Sandars (1980)
Oxford (1984)	-9.3 ± 1.5	-17	Novikov et al. (1976)
Moscow (1984)	-7.8 ± 1.8	-18.8	Barkov (1980a)
		-10.5	(Martensson et al. 1981)
Bi 876 nm			
Seattle (1981)	-10.4 ± 1.7	-8	Martensson et al. (1981)
		-11	Sandars (1980)
		-13	Novikov et al. (1976)

Source: Bouchiat and Pottier (1984).

in the experimental apparatus and carried out more thorough investigations of possible sources of systematic error. In 1981, Bucksbaum and collaborators (1981a,b) reported a value $\delta = (+2.8 \, ^{+1.0}_{-0.9}) \times 10^{-3}$ in comparison with the theoretical value $(+2.1 \pm 0.7) \times 10^{-3}$. The change in the theoretical value of δ was caused by a change in the experimentally measured value of $\sin^2\theta_W$ from 0.25 to 0.23.

Parity-nonconserving optical rotation has also been observed in lead by the Washington group (Emmons et al., 1983). Their experimental value of $R = (-9.9 \pm 2.5) \times 10^{-8}$ agrees, to within the uncertainties of the measurement and the atomic-theory calculation, with the theoretical prediction of $R = -13 \times 10^{-8}$. A series of measurements has also been done on cesium. As early as 1974, even before the existence of the weak neutral currents predicted by W-S theory had been established, Bouchiat and Bouchiat (1974) had calculated the expected effect of such neutral currents in atomic parity–violation experiments. They had found that the effect would be enhanced in heavy atoms: "going from hydrogen to cesium, one gets an enhancement of the order of 10^6" (p. 112). The first experimental result on cesium was reported by Bouchiat and collaborators (1982). They found that the parity-nonconserving parameter $\text{Im}(E_1^{PNC}/\beta)\text{exp} = (-1.34 \pm 0.22 \pm 0.11)$ mV/cm, where the theoretical value was (-1.73 ± 0.07) mV/cm. They concluded, "In view of the experimental and theoretical uncertainties, this is quite consistent with the measured value" (p. 369). This measurement was on a $\Delta F = 0$ hyperfine transition. A second paper (Bouchiat et al., 1984) reported a measurement on a $\Delta F = 1$ transition in cesium and found $Im(E_1^{PNC}/\beta) = (-1.78 \pm 0.26 \pm 0.12)$ mV/cm. "Within the quoted uncertainties, the two results clearly agree, so the two measurements successfully cross-check one another. It is then fair to combine them, which yields $Im\ E_1^{PNC}/\beta = -1.56 \pm 0.17 \pm 0.12$ mV/cm" (p. 467). The theoretical value had changed slightly to $-1.61 \pm 0.07 \pm 0.20$ mV/cm, so theory and experiment were in agreement.

A recent experiment on cesium has been performed by Carl Wieman and his collaborators (Gilbert et al., 1985; Gilbert and Wieman, 1986; Wieman et al., 1987). They found $Im\ E_1^{PNC}/\beta = -1.65 \pm 0.13$ mV/cm, in good agreement with the previous measurement by Bouchiat and collaborators and with theoretical prediction, discussed above. This was the first atomic parity–violation experiment to obtain an uncertainty $<10\%$. The experimental checks were extensive: They included four independent spatial reversals of experimental conditions to identify the parity-non-conserving signal when, in principle, only two are required to resolve the effect. This reduced the potential systematic error because nearly all the factors that can affect the transition rate are correlated with, at most, one of these reversals. Other possible sources of systematic error were identified and their possible effects measured in auxiliary experiments. The experimenters also introduced known nonreversing fields, misalignments, and other confounding factors. The measured effect of these interventions agreed with their calculations of these effects and also indicated that these effects were small compared with the parity-violating signal. Their analysis of their data over time scales from minutes to days also indicated that the distribution of their measured values of $Im\ E_1^{PNC}/\beta$ was completely statistical, and that time-dependent systematic effects were small.

It is fair to say that the current situation with respect to atomic parity–violation experiments and the W-S theory is that the preponderance of evidence favors the theory. The later experiments, which eliminated various sources of background and systematic uncertainty, are more credible than the earliest attempts to measure atomic parity violation. No one knows with certainty why those early results were wrong. Nevertheless, since those early experiments were performed, physicists have found new sources of systematic error that were not dealt with in the early experiments. The redesign of the apparatus has, in many cases, precluded testing whether these effects were significant in the older apparatus. Although it cannot be claimed with certainty that these effects account for the earlier, presumably incorrect, results, there are reasonable grounds for believing that the later results are more accurate. The consistency of the later measurements, especially those made independently by different groups, enhances that belief.

The choice between the early atomic parity–violation results and the SLAC E122 result was determined by evidential weight based on epistemological criteria. The discord between the various atomic parity–violation results was resolved by both the greater credibility of the later results, again based on epistemological criteria, and by a preponderance of evidence.

Conclusion

I began this book by describing two problems for my view that experimental evidence can provide the legitimate basis for scientific knowledge: (1) the question of selectivity, in either data or the analysis of that data; and (2) the resolution of discordant results. Selectivity, and the associated problem of possible experimenter bias, casts doubt on the validity or correctness of an experimental result. It is, as we have seen, often associated with the second problem, the resolution of discordant results. If, as is often the case, experimental results disagree, how can scientific knowledge be based on such results? Although a consensus is usually achieved within a reasonable time, I believe that one must demonstrate that the methods by which such resolution is achieved provide grounds for scientific knowledge—in other words, that they are based on epistemological and methodological criteria.

Selectivity

In Part I, I discussed several types of selection criteria that have been applied to data and to analysis procedures. The cuts have ranged from straightforward and legitimate (as in Millikan's exclusion of data obtained when he was not sure that his experimental apparatus was working properly[1]) to problematic cuts that stemmed from very complex analysis tuning (as in the case of the proposed low-mass electron-positron states, in which the results were an artifact of the cuts). It seems clear that there is no single solution to the problem of determining whether an experimental result is an artifact created by the cuts: What may work in one case may not work in another. There are, however, some general strategies to answer that question.

Consider, for example, robustness. This is an important method of demonstrating the correctness of an experimental result, and for dealing with the problem of cuts. It was, in fact, used in each of the episodes discussed in this book. In the experiment to measure the K_{e2}^+ branching ratio, for example, the experimenters varied both the range cut and the track-matching criterion over reasonable intervals and showed that the branching ratio found was robust under those variations. In the cases of gravity waves and the 17-keV neutrino, robustness again played an important role. In the gravity wave episode, Weber's critics used their own preferred analysis algorithm as well as Weber's nonlinear algorithm and showed that they still found no gravity-wave signal. This was one of the arguments that favored the critics' results over Weber's result. Similarly, in the case of the 17-keV neutrino, several experimenters used both a wide and a narrow energy range in their analysis and demonstrated that their conclusions did not change. In the decisive experiments that showed that the 17-keV neutrino did not exist, the experimenters demonstrated that the choice of analysis procedure was not a problem in their experiments. We also saw how the apparent failure to use robustness as a criterion led to misinterpretation of an artifact of data analysis as a real effect in an independent reanalysis of another researcher's data.

Robustness did not, however, provide an unambiguous solution to the problem in the episode of the low-mass electron-positron states. This was because the results obtained from various experiments, although similar, seemed to be extremely sensitive to a variety of experimental conditions, including time of flight, bombarding energy, scattering angle, and the relative values of the electron and positron energies. Varying these conditions could make the effects vary or disappear: The results lacked robustness. Were the variations a real sensitivity to the conditions or were they artifacts? There are, after all, many phenomena in science which exhibit such sensitivity. Experimenters thought that this sensitivity might pertain in the heavy-ion collisions that produced the low-mass electron-positron states. In this episode, more careful analysis subsequently showed that tuning the cuts could produce the results initially obtained. The results were merely magnifications of statistical fluctuations produced by tuning the selection criteria.

Nevertheless, because similar results were obtained in several experiments using different detectors, different projectile and target nuclei, and at similar (although not identical) energies, the results carried sufficient credibility to encourage further investigation of these heavy-ion collisions. This raises the interesting question of how similar two experiments must be to count as replications and how close experimental results must be to count as confirmations. A large number of possible low-mass electron-positron states were found in the different experiments (see Figure 5.7). The experimenters interpreted the results as evidence for three such states and used the fact that the results were obtained in "different" experiments to support the existence of the states.[2] In retrospect, they were wrong. There is no easy solution to the problem of what constitutes confirmation or replication of an experimental result. How similar the conditions or effects must be can be decided only on a case-by-case basis.

Replication, another form of robustness, also plays an important role in guarding against artifacts created by cuts. In the case of Millikan, unlike the other episodes discussed, both his data exclusion and his varying analysis procedures were private and thus unavailable to the scientific community. Here the robustness of the value of the charge on the electron, obtained in both similar and different experiments, argued for the correctness of Millikan's result and acted as a safeguard against his selectivity. This replication is usually the case in experiments with important theoretical implications. For example, in the discovery of the existence of the intermediate vector boson there were two experiments, UA1 and UA2, each of which demonstrated the existence of the particle. In the case of the SLAC E122 experiment that demonstrated the existence of parity violation in electroweak interactions, the fact that only a single experiment was performed made the epistemological arguments in support of the correctness of the result crucial. The experiment has not been replicated, but the care with which it was done and analyzed has persuaded the physics community that the result is correct (see Chapter 10 for details). The failure, however, to reproduce the low-mass electron-positron effects in Bhabha scattering, a different physical system, but one in which the same effects were expected, cast doubt on the results.[3] There are, however, episodes in which incorrect results have been replicated.

Replication is not a guarantee of the correctness of an experimental result (see Galison, 1987, Chapter 2;[4] and Chapter 5 of this book).

Demonstrating that cuts could create the observed effect also played a major role in several episodes discussed in this book. Thus, Kafka, analyzing his own data and varying his threshold criterion, showed that he could create an apparent gravity-wave signal. The same effect was demonstrated by Levine and Garwin using a computer simulation. In the episode of the 17-keV neutrino, Bonvicini showed, also by means of a Monte Carlo calculation, that analysis cuts combined with limited statistics could produce effects that might mask or mimic the presence of the proposed particle. It should be emphasized, however, that demonstrating that an effect can be produced by applying selection criteria can only cast doubt on an experimental result. It cannot demonstrate that the result is incorrect. In the cases of gravity waves and the low-mass electron-positron states, other arguments were both needed and provided. Conversely, arguing that the applied cuts could not create the observed effect (as was the case for the K_{e2}^+ branching-ratio experiment) increased confidence in the result.

Sometimes one can argue that an experimental result is not an artifact by the use of a surrogate signal. Detection of the surrogate signal argues that the experimental apparatus and the analysis procedure are working properly. This was the case in the episodes of both gravity waves and the 17-keV neutrino. Weber's critics were able to detect a pulse of acoustic energy injected into the antenna that mimicked the effect expected for gravity waves. The Argonne group was able to detect the kink created by the composite spectrum of ^{35}S and ^{14}C, which served as a surrogate for the effect expected for the 17-keV neutrino. Such a procedure tests the proper operation of both the experimental apparatus and the analysis procedure, including the cuts.

Because there is no single algorithm or procedure to guard against results that are artifacts of the selection criteria, should we doubt both experimental results and the science based on those results? I think not. Although, as we have seen, the correctness of results may be difficult to establish, it is not impossible to do so. In each of these episodes discussed, the question of whether the result was an artifact was answered. It would be an error to conclude that because three of the five cases discussed in Part I had results that were artifacts of the selection criteria, that this is typical of experimental results in physics. The episodes were chosen precisely because there were discordant results and the selection criteria were important. The K^+_{e2} branching ratio experiment is the norm, not the exception. Cuts may be ubiquitous, but they are not fatal.

I have also discussed a method that is currently being used by physicists to avoid the problem of selectivity. This is the technique of blind analysis, in which the result of an experiment is kept unknown until the analysis of the data is essentially complete. It is clear from the amount of effort devoted to blind analysis that the problem of selectivity is one that troubles the physics community. I suspect that blind analysis will increasingly be applied to experiments. The technique is not, however, easily applied to all experiments and in those cases in which it cannot be applied, the strategies we discussed above will, no doubt, be used. As we have seen, the problem of selectivity can be, and has been, solved.

The Resolution of Discordant Results

In Part II and in Chapters 2 and 4, I have argued that in six separate episodes in modern physics, the discord between experimental results was resolved by reasoned discussion based on epistemological and methodological criteria. Alan Nelson (1994) has suggested that historical accounts, such as those I have given, are insufficient to establish the superiority of a rational or reasonable account over a constructivist one. He states, in discussing the atomic parity–violation episode discussed in Chapter 10:

Franklin does a lovely job of showing, once all the actual evidence was in, the Standard Model could have been regarded as more strongly supported than the hybrids. But, the constructivist should reply that this is yet another exercise in retrospective rationalism. *After* scientists make a choice in a case like this, they naturally go on to *construct* the kind of evidence that supports their choice. In a possible world where scientists preferred hybrid models, experiments would have been tuned differently, etc. so that the constructed evidence would have rationally supported a hybrid model. A Franklin counterpart in that possible world would be arguing that hybrid theories were chosen on rationalist grounds! (Nelson, 1994, p. 546)

Nelson has placed the cart before the horse. Scientists decide what the valid experimental evidence is and then make their theory choice, not vice versa. Scientists have an interest in producing scientific knowledge, as well as a career interest in being correct, and such a procedure is far more likely to produce a correct choice. Without evidence as a guide, how are scientists supposed to make such a choice? They might just as well flip a coin.

I believe that Nelson also overestimates the plasticity of nature and experimental practice. He is, of course, correct that experimenters often modify their practice as they perform the experiment and analyze their data to produce a result. Not all such possible procedures can, however, be justified. For example, a scientist who excluded all those experimental runs whose results did not agree with his preferred theory would not be credible. If that fraud became known, the scientist would be ostracized—not everything goes. In addition, I believe that Nelson overstates just how much one can change results using legitimate procedures. It would, for example, require dramatic and unjustifiable modifications of apparatus and analysis procedures to demonstrate that objects whose density is greater than that of air fall up when released.

Collins and Pickering have offered constructivist accounts of two of the episodes I have discussed, atomic parity violation and gravity waves. I believe their accounts are incorrect.[5] In his most recent comments on the atomic parity–violation episode, Pickering (1991) argues that my view that the decision of the physics community to accept the W-S theory on the basis of reasonable evidence fails because there were too many reasons for that decision. He presents four alternative scenarios, which he regards as equally reasonable resolutions of the problem. None of these alternatives was actually chosen to resolve the discord, leading Pickering to argue that because reason was unable to decide the issue, the scientists allowed the prospect of future research opportunities to influence their resolution. Three of the alternatives proposed by Pickering involve questions about the evaluation of experimental evidence:

1. The physics community might have decided that the atomic parity–violation results of Washington-Oxford were wrong and were therefore excluded.[6]

2. The physics community might have lumped the atomic parity–violation results together with those of SLAC E122 and concluded that they neutralized each other.[7]

3. The community might have waited until E122 had been replicated before making a decision.[8]

Pickering regards these alternatives as being as reasonable as accepting the SLAC E122 results and awaiting further work on atomic parity violation. He is somewhat alone: None of these alternatives were pursued. He has presented no reasons why they should have been. He has merely asserted that they were equally reasonable. I have argued elsewhere that given the evidential context, they are in fact not equally reasonable alternatives (Franklin, 1993b).

Pickering might also deny that the physics community engaged in an evaluation of the experimental evidence. (See, however, the statements by Dydak and Bouchiat, in Chapter 10, in which they do evaluate the evidence.) However, the atomic parity–violation experiments—including repetitions of the original Washington-Oxford experiments on bismuth—continued through the 1980s and into the 1990s, reaching agreement with the predictions of the W-S theory.[9] If the original Oxford-Washington results were simply regarded as wrong, there seems little reason for experimentation to continue. If, however, judgment was suspended concerning which of the discordant results was correct, then the subsequent experimental work certainly makes sense, and even seems to be required.

Pickering also asks why a theorist might not have attempted to find a variant of electroweak gauge theory that might have reconciled the Washington-Oxford atomic parity results with the positive E122 result. (What such a theorist was supposed to do with the supportive atomic parity results of Berkeley and of Novosibirsk is never mentioned.) "But though it is true that E122 analyzed their data in a way that displayed the improbability [6×10^{-4}] of a particular class of variant gauge theories, the so-called 'hybrid models,' I do not believe that it would have been impossible to devise yet more variants" (Pickering 1991, p. 462). Pickering notes that open-ended recipes for constructing such variants had been written down as early as 1972. I agree that it would certainly have been possible to do so, but one may ask whether scientists would consider this a productive use of their time. If the scientists agree with my view that one had reliable evidence (E122 and others) that supported the W-S theory and a set of conflicting and uncertain results from atomic parity–violation experiments that gave an equivocal answer in support of the W-S theory, what reason would they have to invent an alternative?

Constructivists like to claim that they are only describing scientific practice and not making judgments. Both Pickering and Collins seem to ignore this dictum—in fact, they substitute their judgment for that of the scientific community. In the case of gravity waves, Collins (1985) has stated:

Under these circumstances it is not obvious how the credibility of the high flux case [Weber's results] fell so low. In fact, it was not the single uncriticized experiment that was decisive. . . . Obviously the sheer weight of negative opinion was a factor, but given the tractability, as it were, of all the negative evidence, it did not *have* to add up so decisively. There was a way of assembling the evidence, noting the flaws in each grain, such that outright rejection of the high flux claim was not the necessary inference. (p. 91)

Collins also presents alternatives that were plausible to him, but not to scientists working in the field. As I have shown, there were good reasons for rejecting Weber's results.

I have previously argued that it is insufficient for constructivists to merely claim that things *could* have been different in a particular episode involving discordant results, but that they must argue either that things *should* have been different or that other criteria were used. Nick Rasmussen (pers. comm.) has suggested that I am holding constructivists to an impossibly high standard. He says that examination of the published record will never show scientists making a decision that goes against experimental evidence.[10] This is because scientists always give reasons for their decision that will appeal to and persuade the scientific community. Why such reasons are persuasive to members of the scientific community is not discussed by constructivists. Rasmussen states that constructivists will never be able to show that the situation was different or that it should have been different, using such evidence.

I disagree. Rasmussen's view requires that we believe that scientists do not give their "real" arguments—that they are presenting only those arguments that will persuade their fellow scientists. There is, however, no evidence that the public and private arguments are different. In one case in which I have been able to examine, both the private e-mail correspondence between the proposers of the Fifth Force and their published response to criticisms of the proposal, there was no such difference (Franklin 1993a, pp. 35–48). There are also other sources available to the historian of science: Notebooks, letters, e-mail, and the like could all show that the public and private reasons differ. In fact, Collins (1985) has claimed, in his study of gravity waves, that the public and private reasons are different. Based on interviews with scientists, he concluded that the community need not have rejected Weber's results. Collins's claim disagrees with the published discussion (at GR7) mentioned in Chapter 2. Although individual scientists may find fault with particular bits of evidence, that does not mean that the overall decision, based on all of the evidence, is unreasonable.

If, as I strongly believe, scientific knowledge is grounded in experimental evidence, then we must have good reasons for belief in experimental results. In the Introduction, I outlined an epistemology of experiment, a set of strategies that can be—and is—legitimately used to argue for the correctness of an experimental result. In an ideal world, these strategies would always be applied properly and all experimental results would be correct. As we have seen, however, in the real world experiments often give discordant results. I have argued that the discord between experimental results is resolved by reasoned discussion based on epistemological and methodological criteria.

I have also discussed selectivity, another possible problem in determining the correctness of experimental results. Because selection criteria are always applied to either experimental data or to the procedures used to analyze that data, answering the question of whether a result is correct or is an artifact produced by the application of those criteria is of crucial importance. We must answer that question before we can depend on experimental results as the basis of scientific knowledge. In Part I, I argued that the problem of selectivity can be solved.

One point that should be clear from the episodes discussed is that there is no instant rationality in science. Problems of selectivity and discord may take some time to resolve. The episode of the existence of the 17-keV neutrino lasted eight years; that of the Fifth Force, four years; and that of gravity waves, seven years (at least for resolution of the initial controversy between Weber and his critics). The latter is still a subject of current research. The question of the correctness of the LSND neutrino result is still unanswered five years after the initial publication of the result. Although such questions might take some time to answer, they are eventually answered, and those answers are based on experimental evidence and on reasoned and critical discussion.

I have argued that we have good reasons to believe in experimental results and that the problems of selectivity and discord can both be solved. It follows, then, that we may reasonably use experimental evidence as the basis of scientific knowledge.

Notes

Introduction

1. To be fair, Collins claims that this statement is a methodological prescription. The sociologist of science should behave as if "the natural world has a small or non-existent role in the construction of scientific knowledge." There is no such qualification in the passage in which this quotation appears. The qualification appears elsewhere in the essay.

2. Barnes view is a rather strong statement of what is known as "the underdetermination of theory by evidence," discussed in note 25.

3. In later work, Pickering does seem to allow a role for the natural world in the production of experimental results and in investigating theories, but it is not a very important role. This is discussed in detail below.

4. By valid, I mean that the experimental result has been argued for in the correct way, by use of epistemological strategies such as those discussed below.

5. One must be careful here to distinguish between an argument for the existence of an entity and that for the validity of an experimental result. The problem arises here because the result is the existence of the dense bodies.

6. See Franklin (1986, Chapter 6; 1990, Chapter 6) and Franklin and Howson (1984, 1988) for details of these strategies, along with a discussion of how they fit into a Bayesian philosophy of science.

7. Harry Collins (1985) argues that calibration cannot be used to validate experimental results. "The use of calibration depends on the assumption of near identity of effect between the surrogate signal and the unknown signal that is to be measured (detected) with the instrument" (p. 105). Collins further argues that the adequacy of the surrogate signal is not usually questioned by scientists and that calibration can only be performed provided that this assumption is not questioned too deeply. I have argued elsewhere in detail that Collins is wrong (Franklin, 1997a; 1999, pp. 237–72). The question of the adequacy of the surrogate signal is one that experimental physicists consider carefully, and they offer arguments for that adequacy. In many cases the adequacy of the calibration is clear and obvious. There are also, as we shall see, instances that involve

discordant results or other controversies, in which the question of calibration may be both difficult to answer and of paramount importance. This is particularly true when a new type of experimental apparatus is used to search for a hitherto unobserved phenomenon. The episode that Collins uses to support his view of calibration, that of the early attempts to detect gravity waves, is just such an instance. As discussed in detail in Chapter 2, in this case other arguments were both needed and provided.

8. As Holmes remarked to Watson, "How often have I said to you that when you have eliminated the impossible, whatever remains, *however improbable,* must be the truth" (Conan Doyle, 1967, p. 638).

9. Kepler's Third Law was not available when Galileo made his observations, but it is an argument that could have been used later.

10. This change is usually attributed to Arthur Rosenfeld of the University of California at Berkeley. The attribution may be apocryphal, but the high-energy physics community did change its criterion for the existence of a new particle.

11. It might be useful here to distinguish between the theory of the apparatus and the theory of the phenomenon. Ackermann is talking primarily about the latter. It may not always be possible to separate these two theories. The analysis of the data obtained from an instrument may very well involve the theory of the phenomenon, but that does not necessarily cast doubt on the validity of the experimental result.

12. For another episode in which the elimination of background was crucial, see the discussion of the measurement of the K_{e2}^{+} branching ratio in Franklin (1990, pp. 115–31) and Chapter 1.

13. Galison's previous position was that experimenters within the two traditions had a preference for certain types of argument. I agree. His more recent view makes this distinction more rigid. The fact that different groups of experimenters use different strategies or arguments in support of the credibility of their experimental results does not raise any problems for my epistemology of experiment. As I noted earlier, none of the strategies is necessary and it is not surprising that experiments using a certain type of apparatus use only certain strategies. What is more problematic is if experimenters using different types of apparatus belong to different language groups and therefore cannot understand each others' epistemological arguments. As discussed in this Introduction, neither Staley nor I think that this is the case.

14. Galison constructs an elaborate structure based on the rigid dichotomy between the two traditions and their different languages and forms of argument. He notes that they have now merged into a hybrid tradition, in which the detectors combine many of the best features of the two traditions. The detectors thus provide very detailed information about large numbers of events. The data are recorded electronically and are used to construct computer images of the events that are, in Kent Staley's (1999) term, "visually isomorphic" to the event. The large number of events allows the use of statistical techniques. Because the two traditions also have different languages, their merger necessitates, in Galison's view, "trading zones," in which communication by "pidgin" and "creole," rather than in a common language takes place. I discuss some of my disagreement with this scheme below.

15. This will also be discussed in the case studies presented in this book.

16. One might regard this episode as a golden "golden" event. Staley has some reservations concerning this. He points out that Anderson (1933) not only includes photographs of four events but also states that:

out of a group of 1300 photographs of cosmic-ray tracks 15 of these show positive particles penetrating the lead, none of which can be ascribed to particles with a mass as large as that of a proton, thus establishing the existence of positive particles of unit charge and of mass small compared with that of the proton. (p. 493)

17. Anderson named the particle the positron.

18. It is, in fact, possible for a particle to increase its energy in passing through matter. It is, however, extremely unlikely. Note, for example the occasional increase in energy of particle undergoing Brownian motion, which does not violate the conservation of energy. Anderson does not, in fact, invoke the conservation of energy. He seems to regard the presumed effect as very unlikely. "We also discarded as completely untenable the assumption of an electron of 20 million volts entering the lead on one side and coming out with an energy of 60 million volts on the other side" (Anderson 1933, p. 491).

19. For some of the technical details of this analysis, see Staley (1999, pp. 203–8).

20. I myself searched for the elusive "bump" on several occasions during that period.

21. Although, for the reasons discussed here, this was not a golden event, I believe that it does provide a counterexample to Galison's view that such events cannot occur within the logic tradition.

22. There is also the possibility that the initial candidate was caused by someone giving the apparatus a good whack. Why this should have been done, or why it should have resulted in a signal of exactly the right size for a monopole, are further mysteries.

23. Galison's discussion contains no explicit quantitative statistics, but is clearly probabilistic. See also Staley (1999, pp. 208–13).

24. I believe that the evidence presented by Staley (and, interestingly, by Galison himself) argues strongly that the two communities are not linguistically and epistemically distinct. They share an underlying statistical method. This is not to say, of course, that no other strategies are used, but only that there is sufficient shared method to allow for easy communication. My view is further supported by the fact that a significant number of high-energy physicists have worked easily in both traditions. At least three Nobel Prize winners, Martin Perl, Melvin Schwartz, and Jack Steinberger, are included in that group. More personally, my colleague Uriel Nauenberg and I have both done both bubble-chamber and spark-chamber experiments. Uriel's initial training was in bubble-chamber techniques, whereas I initially used spark chambers. I later worked on bubble-chamber experiments. Although I had to learn new material, it certainly was not the equivalent of learning a new language. Nor did the forms of argument change. The "bump hunting" experiments I worked on used statistical methods. During the 1970s, Uriel and I both worked on a wire-chamber experiment to measure the π^0 energy spectrum in K^0 meson decay. We had no difficulty in communicating. I doubt that we are unique.

25. The Duhem-Quine problem is related to what is called the underdetermination of theory by evidence: One can always construct an alternative theory that explains a given set of data. This is trivially correct. Quantum mechanics predicts the Balmer series

in hydrogen, but so does quantum mechanics conjoined with the statement "The moon is made of green cheese." (For a discussion of this so-called "tacking paradox," see Howson and Franklin [1986].) The question that arises, however, is whether such an alternative theory is in any way significantly different from the theory currently on offer, and whether it is physically interesting. For further discussion of this issue, see Franklin (1988), and for a particular example of a pragmatic solution to the Duhem-Quine problem (in which not all the logically possible alternatives were considered, but only those that were thought to be physically interesting and plausible), see Franklin (1986, Chapter 3).

26. Collins offers two arguments concerning the difficulty, if not the virtual impossibility of replication. The first is philosophical. What does it mean to replicate an experiment? In what way is the replication similar to the original experiment? A rough and ready answer is that the replication measures the same physical quantity. Whether it in fact does so can, I believe, be argued for on reasonable grounds, as discussed earlier.

Collins's second argument is pragmatic: In practice it is often difficult to get an experimental apparatus, even one known to be similar to another, to work properly. Collins illustrates this with his account of Harrison's attempts to construct two versions of a transverse excited atmospheric (TEA) laser (Collins, 1985, pp. 51–78). Despite the fact that Harrison had previous experience with such lasers, and had excellent contacts with experts in the field, he had great difficulty in building the lasers. Hence the difficulty of replication.

Ultimately Harrison found errors in his apparatus, and once these were corrected, the lasers operated properly. As Collins (1985) admits:

> in the case of the TEA laser the circle was readily broken. The ability of the laser to vaporize concrete, or whatever, comprised a universally agreed criterion of experimental quality. There was never any doubt that the laser ought to be able to work and never any doubt about when one was working and when it was not. (p. 84)

Although Collins seems to regard Harrison's problems with replication as casting light on the episode of gravity waves, as support for the experimenters' regress, and as casting doubt on experimental evidence in general, it really does no such thing. As Collins concedes, the replication was clearly demonstrable.

27. In more detailed discussions of this episode, Franklin (1994, 1997a), I argued that the gravity wave experiment is not at all typical of physics experiments. In most experiments, as illustrated in those essays, the adequacy of the surrogate signal used in the calibration of the experimental apparatus is clear and unproblematical. In cases where it is questionable considerable effort is devoted to establishing the adequacy of that surrogate signal. Although Collins has chosen an atypical example I believe that the questions he raises about calibration in general and about this particular episode of gravity wave experiments should be, and can be, answered.

28. Morpurgo did observe integral charges in his very small initial data sample. When he took further data, the continuous values appeared.

29. Note that Morpurgo does not agree with Pickering's interpretation of this episode:

> I want to make it clear, however, that Professor Morpurgo offers no endorsement of what follows. He wrote of my earliest account of his work that "[w]e certainly appreciate the

intention of the study of Dr. Pickering; however we disagree with many of his statements and this continues to be the case for the present chapter. . . . More generally, Professor Morpurgo has asked me to state that he "does not share many aspects of my general view of the interrelationship between theory and experiment." (Pickering 1995, p. 69)

30. One could conceivably construct a scenario in which fractional charges exist on the niobium spheres used by Fairbank but not on the graphite and iron used by Morpurgo. Although this seems highly improbable, it could happen. When parity-violating (violations of left-right symmetry) were seen in high-energy neutrino interactions, but not in low-energy atomic interactions, theorists did construct models that accommodated both results (see Chapter 10).

31. For a philosophical discussion of the structure of scientific papers, see Lipton (1998), Suppe (1998a,b), and Franklin and Howson (1998).

32. This is a rather different categorization than the old rationalist-empiricist distinction of standard philosophy. In that classification those who, like Descartes, believed that one could acquire knowledge of the world by pure thought would be the rationalists, whereas those who thought that experience was the sole source of such knowledge would be the empiricists (see Ackermann [1985, Chapter 1]). I, myself, would prefer to be described as a rational empiricist. I would actually prefer the term logical empiricist to acknowledge my intellectual debt, but that term is in current disrepute. As John Passmore (1972) wrote in the *Encyclopedia of Philosophy*, "Logical positivism [or logical empiricism] is dead, or as dead as a philosophical movement ever becomes" (p. 56). I disagree with Passmore's assessment, but that is a subject for discussion elsewhere (see, however, Creath [1995]).

33. Hacking seems to believe that an alternative physics would necessarily have different standards of success. I do not think this is correct. One can imagine an alternative physics that would be just as successful by the same standards.

34. Although it is logically possible that someone might come up with a view of the world that does not involve light or its speed, I think this extremely improbable. If light were a part of anyone's furniture of the world, I believe its velocity would be the same as it is now, at least within experimental uncertainty.

35. For details of this episode, see Franklin (1986, Chapter 3).

36. Note that the physics community was accepting a result that seemed to refute a strongly supported and well-established symmetry law, and we can see that it led to large amounts of both theoretical and experimental work.

37. Van Fraassen defines observable as "detectable with unaided human senses." I disagree. See Franklin (2000a, pp. 313–18).

38. Interestingly, the title of the second edition is *Laboratory Life: The Construction of Scientific Facts*. The "social" has disappeared.

39. For an extended discussion of this issue, see Franklin (2000a). For an opposing view, see Dancoff (1952).

40. For differing accounts of various episodes see: weak neutral currents (Pickering, 1984b; Galison, 1987); solar neutrinos (Shapere, 1982; Pinch,1986; Franklin, 2000a); atomic parity–violation experiments (Pickering, 1984a; Franklin, 1990, Chapter 8; Ackermann, 1991; Franklin, 1991; Lynch, 1991; Pickering, 1991; Franklin, 1993b); and early searches for gravity waves (Collins, 1985; Collins, 1994; Franklin, 1994, 1998). For a further discussion of these issues, see also Koertge (1998).

41. On this scale, Hacking gives Thomas Kuhn a score of 5, 5, and 5. I suspect that Kuhn, as well as several of his commentators, would strongly disagree.

42. I know that this view results in some philosophical problems and that other philosophers of science call knowledge "justified, true belief." I distinguish between knowledge and truth. Thus, I believe that Newton's laws of motion and his law of universal gravitation were knowledge, certainly for the 18th and 19th centuries, although we currently regard them as false.

Part I. Selectivity and the Production of Experimental Results

1. I will not deal here with the selectivity that is built into an experimental apparatus.

2. One should distinguish between experimental data and an experimental result: They are usually different. (See Chapter 1, note 3). What I mean by "analysis procedures" are those processes that transform data into an experimental result. These processes may involve computer analysis and simulation, making cuts on the data, and other procedures. This distinction will be illustrated in the episodes discussed in Chapters 1–5.

3. By valid, I mean that the experimental result has been argued for in the correct way, using epistemological strategies discussed in the Introduction.

4. This is the issue of calibration. Some critics have questioned the use of calibration to validate a result. For a fuller discussion see Franklin (1997a).

1. Measurement of the K_{e2}^+ Branching Ratio

1. Strangeness is a property of elementary particles. It is conserved in strong and electromagnetic interactions, but not in weak interactions. For details see Franklin (1986, Chapter 3).

2. How far a charged particle will travel in matter before stopping depends on its velocity. For a momentum-selected beam, kaons—which are lighter than protons but heavier than pions—will have a higher velocity than that for the protons in the beam and a lower velocity than that for the pions. Thus the kaons will travel through more material than protons but have a shorter range than do pions. In the experiment, enough copper was placed in the beam in front of the kaon-stopping region to remove all of the protons, whereas the pions that passed through the stopping region were counted in Cu_4 and vetoed. (If a beam particle was counted in Cu_4, the spark chambers were not triggered.) The difference in velocity also resulted in a difference in the time of flight of the particles from the production target to the stopping region.

3. These events are not raw data. The data for the experiment consisted of spark-chamber and oscilloscope photographs along with scaler readings. The sparks were fitted to a trajectory, which—combined with the known magnetic field—allows a determination of the decay particle's momentum. These momenta are plotted in Figure 1.2.

2. Early Attempts to Detect Gravity Waves

1. The ratio of the gravitational force between the electron and the proton in the hydrogen atom to the electrical force between them is 4.38×10^{-40}, a small number indeed.

2. This device is often referred to as a "Weber bar."

3. Given any such threshold, there is a finite probability that a noise pulse will be larger than that threshold. The point is to show that there are pulses in excess of the number expected statistically.

4. It was also pointed out that if there were a real sidereal effect, it should have had a 12-hour period. Passage through the earth should have had very little effect on the probability of detecting the gravity waves.

5. Note here the repetition of experiments measuring an important physical quantity. This will also be significant in the next chapter.

6. In this discussion, I have relied primarily on a panel discussion on gravity waves that took place at the Seventh International Conference on General Relativity and Gravitation (GR7), Tel-Aviv University, June 23–28, 1974. The panel included Weber and three of his critics, Tyson, Kafka, and Drever, and included papers presented by the four scientists and discussion, criticism, and questions. It includes almost all of the important and relevant arguments concerning the discordant results. The proceedings were published as Shaviv and Rosen (1975). Unless otherwise indicated, all quotations in this chapter are from Shaviv and Rosen (1975). I give the author and the page numbers in the text.

7. Drever summarized the situation in June 1974 as follows:

Perhaps I might just express a personal opinion on the situation because you have heard about Joseph Weber's experiments getting positive results, you have heard about three other experiments getting negative results and there are others too getting negative results, and what does this all mean? Now, at its face value there is obviously a strong discrepancy but I think it is worth trying hard to see if there is any way to fit all of these apparently discordant results together. I have thought about this very hard, and my conclusion is that in any one of these experiments relating to Joe's one, there is always a loophole. It is a different loophole from one experiment to the next. In the case of our own experiments, for example, they are not very sensitive for long pulses. In the case of the experiments described by Peter Kafka and Tony Tyson, they used a slightly different algorithm which you would expect to be the most sensitive, but it is only the most sensitive for a certain kind of waveform. In fact, the most probable waveforms. But you can, if you try very hard, invent artificial waveforms for which this algorithm is not quite so sensitive. So it is not beyond the bounds of possibility that the gravitational waves have that particular kind of waveform. However, our own experiment would detect that type of waveform; in fact, as efficiently as it would the more usually expected ones, so I think we close that loophole. I think that when you put all these different experiments together, because they are different, most loopholes are closed. It becomes rather difficult now, I think, to try and find a consistent answer. But still not impossible, in my opinion. One cannot reach a really definite conclusion, but it is rather difficult, I think to understand how all the experimental data can fit together. (pp. 287–88)

8. I have been unable to find the published proceedings of this conference. Richard Garwin (private communication) has informed me that these proceedings were never published.

9. As Weber answered, the Maryland group had presented data showing no positive coincidence excess at GR7. Garwin was not, however, at that meeting, and the proceedings were not published until after Garwin's 1974 letter appeared.

10. I discuss the legitimate use of such simulations in Chapter 6.

3. Millikan's Measurement of the Charge of the Electron

1. For a discussion of some of these early experiments, see Millikan (1917, Chapter III).

2. The subscripts indicate terminal velocities without (v_g) and with (v_f) the field, respectively. Now m can be replaced by a using $m = 4/3\pi a^3(\sigma - \rho)$, σ and ρ being the densities of oil and air, respectively; a can be done away with in favor of μ using Stokes' law; and the ratios of distance d, to times of all and rise, t_g and t_f can be substituted for the velocities. Millikan did not make all of these substitutions. He left a factor of v_g in his final formula, presumably for ease of calculation.

3. Millikan was too optimistic. Such instrumental effects as electric field inhomogeneities and space-charge effects limited the accuracy of his measurements to drops with charges less than about $30e$. See Fairbank Jr. and Franklin (1982) and the discussion in note 12.

4. This is a very small uncertainty. Millikan estimated it using the statistical uncertainty in his final value for e. He did not include any uncertainty caused by systematic effects.

5. Millikan's value for e differs from the modern value $e = (4.80320420 \pm 0.00000019) \times 10^{-10}$ esu. This difference is due, in large part, to a difference between the modern value for the viscosity of air and the one that Millikan used.

6. My work here is based on Millikan's notebooks at the California Institute of Technology. For details of my recalculation of Millikan's data, see Franklin (1981).

7. Daniel Siegel raises the same question. "Millikan was in this sense choosing data according to his presuppositions, and then using those data to support his presuppositions"(Siegel, 1979, p. 476). As discussed in Franklin (1981), and in this chapter, I disagree with Siegel's statement.

8. Recall that Millikan modified Stokes's law, substituting $K/(1 + b/pa)$ for K to take into account the particulate nature of air in the experiment. The parameter b was empirically determined from the entire data set and has an uncertainty. In addition, this was a first-order approximation to Stokes's law. Other terms in the approximation may have been important. This was the case for the 12 drops, discussed below, that Millikan did not publish because they seemed to require a second-order correction to Stokes's law.

9. The quotations are from Millikan's notebooks.

10. Interestingly, the value for e that one finds from these 68 excluded events was $e = (4.75 \pm 0.01) \times 10^{-10}$ esu. This was, in fact, more precise than any other measurement available at the time. The data were, however, untrustworthy.

11. I attempted, without success, to calculate a second-order correction to Stokes's law for these 12 drops. I found no consistent way to do so.

12. The second drop of 16 April 1912 is quite anomalous. (The data sheet for this drop is shown in Figure 3.2). It is also quite worrisome because it is among Millikan's most consistent measurements. Not only are the two methods of calculating e internally consistent, but they agree with each other very well. Millikan liked it: "Publish. Fine for showing two methods of getting v." My own calculation of e for this event gives a value $e = 2.810 \times 10^{-10}$ esu, or approximately $0.6e$. Millikan knew this. Note the comment,

"Won't work" in the lower right-hand corner. There were no obvious experimental difficulties that could explain the anomaly. Millikan remarked, "Something wrong with therm[ometer]," but there is no temperature effect that could by any stretch of the imagination explain a discrepancy of this magnitude. Millikan may have excluded this event to avoid giving Ehrenhaft ammunition in the controversy over the quantization of charge. In retrospect Millikan was correct in excluding this drop. In later work William Fairbank Jr. and I found that Millikan's apparatus gave unreliable charge measurements when the charge on the drop exceeded a value of about 30*e*. This drop had a charge of greater than 50*e*, and the data were quite unreliable (Fairbank Jr. and Franklin, 1982).

13. This group of 19 drops included some of those used in Millikan's final calculation of *e* as well as some of those omitted from the calculation.

14. The effect of Millikan's selectivity was to reduce the statistical uncertainty of his final result very slightly. It had no significant effect on the final value of *e*. (See Table 3.1.) Almost all of the uncertainty in Millikan's final value was due to systematic effects—uncertainty in the distance between the plates, uncertainty in the voltage, and the like. One may wonder why he was so worried about the statistical uncertainty.

15. Although the fact that Weber used a threshold cut was publicly known, the value of that cut, and whether he used a single threshold value was not known. This lessened the credibility of his result.

16. It is rare that an important physical quantity is measured only once. Recall the numerous attempts to replicate Weber's experiment on gravity waves, discussed in Chapter 2. Other examples abound. See, for example, Franklin (1986, Chapters 1–3) and Chapter 5 in this volume on electron-positron states. The only instance I can think of in which an important physical quantity was measured only once is the SLAC E122 experiment. This experiment measured the asymmetry in the scattering of polarized electrons from deuterons, an important prediction of the Weinberg-Slam unified theory of electroweak interactions. Because of its expense and complexity, the experiment was done only once. For that reason the arguments for the validity of its results were extensive. See Franklin (1990, Chapter 8) and Chapter 10 in this volume.

4. The Disappearing Particle

1. For a discussion of the ordinary neutrino, see Franklin (2000a,b).

2. There was also suggestive, although not conclusive, evidence from a third type of experiment, that detecting internal bremsstrahlung in electron capture (IBEC), a form of beta decay. Not all of the IBEC experiments gave positive results. In addition, as discussed below, one of the experiments that convinced the physics community that the 17-keV neutrino did not exist, that of Mortara et al. (1993), used the same type of solid-state detector that Simpson had used.

3. See Franklin (1990, Chapter 1) for details of some early experiments.

4. In a normal beta-decay spectrum the quantity $K = (N(E)/[f(Z,E) (E^2 - 1)^{1/2} E])^{1/2}$ is a linear function of E, the energy of the electron. A plot of that quantity as a function of E, the energy of the decay electron, is called a Kurie plot.

5. Later work, including some by Simpson, reduced the size of the positive effect to approximately 1%.

6. These corrections were extremely important in analyzing the data. See below.

7. Kalbfleisch and Milton (1985) also argued that Simpson's analysis required an incorrect value for the endpoint energy of the tritium spectrum.

8. That these positive results were reported by someone other than Simpson may have given credence to the result in the eyes of the physics community.

9. Bonvicini's work was very important. By showing that a smooth shape-correction factor might either mask or enhance a kink due to a 17-keV neutrino, he cast considerable doubt on the early negative results obtained with magnetic spectrometers. This work was influential in persuading scientists to perform the later, more stringent, experimental tests.

10. In addition, Morrison (1992) showed that Simpson's most persuasive reanalysis of Ohi and collaborators' early negative result was dependent on a statistical fluctuation. Hetherington et al. (1987) had also suggested that this might be a problem.

11. These results were essentially the same as those reported by Kawakami et al. (1992). In his published paper, Bonvicini (1993) agreed with this evaluation.

12. Hime, one of Simpson's collaborators, agreed. "The difficulty remains, however, that an analysis using such a narrow region could mistake statistical fluctuations as a physical effect. The claim of positive effects in these cases [by Simpson] should be taken lightly without a more rigorous treatment of the data" (Hime, 1992, p. 1303).

13. There was yet another problem with the analysis of Ohi and collaborators. As Borge et al. (1986) noted:

> We feel, *in complete agreement with the opinions expressed by J. J. Simpson . . . that the limits on c_2 derived in (the experiments of Ohi et al. (1985) and of (Datar, Baba et al. 1985)) are misleading as the parameters were not fitted again under the assumption of a heavy neutrino; instead the contribution from this was simply added.* (pp. 593–94, emphasis added)

14. Perhaps if Simpson had varied the endpoints of the energy range he used in his reanalysis, he would have avoided his difficulty. As Morrison clearly showed, Simpson's reanalysis of Ohi's data lacked robustness.

15. Magnetic spectrometer experiments required a shape-correction factor, usually of the form $(1 + \alpha E)$, to fit their spectra. This was an important issue in the resolution of the discord. For details, see Franklin (1995a).

16. Bonvicini had reanalyzed many of the early experiments. He showed, using Monte Carlo techniques, that the shape-correction factor needed in the magnetic spectrometer experiments, combined with the limited statistics of those experiments, could mask or mimic the presence of a heavy neutrino. His analysis showed that the negative evidence provided by the early replications of Simpson's experiment was not as strong as had been originally claimed. He did, however, conclude that the experiment of Hetherington et al. (1987) was sufficient to rule out a 3% effect (Bonvicini 1993). Bonvicini's work also influenced the design of later experiments.

17. The upper limit found by the Argonne group was $\sin^2\theta = -0.0004 \pm 0.0008$ (statistical) ± 0.0008 (systematic). This was also far lower than 1%.

18. This is an example, albeit a complex one, of the calibration of an experimental apparatus. For details see Franklin (1997a).

19. Several other negative results were also published at this time. For details see Wietfeldt and Norman (1996).

5. Are There Really Low-Mass Electron-Positron States?

1. As we shall see, these peaks appeared only when certain cuts were applied. In particular, the electrons and positrons were required to have approximately equal energies and to be emitted back to back, exactly what one would expect if they were the decay products of a single state or particle.

2. The EPOS I group was one of those that reported the original effect. The I and II refer to different versions of the experimental apparatus. The membership of the group also changed.

3. Ganz divided his data set into two subsets by using a random-number generator on an event-by-event basis. This guarded against any systematic effects that varied with time.

4. The question of how one should properly estimate such statistical confidence levels when cuts are applied subsequently became an issue. The probability of a six-standard-deviation statistical effect is 2.0×10^{-9}.

5. The signal could also be enhanced relative to the background by making cuts on the heavy-ion scattering angles. See Figure 3 and Figure 32 in Cowan et al. (1987) and their discussion on pp. 185–86.

6. The Moriond Workshops provide a forum for speculative work in physics. Thus from 1987 to 1990 the Fifth Force, a proposed modification of Newton's law of gravity, was extensively discussed at the workshops. For details, see Franklin (1993a).

7. These results were published later as Koenig and collaborators (1989).

8. I am assuming that the center of mass of the produced particle is moving in the same direction as the overall center of mass of the heavy-ion system.

9. The UNILAC at GSI was shut down in 1989 for improvements.

10. Greiner and Reinhardt (1995, p. 218) noted that "The situation may resemble another long-standing experimental puzzle at the beginning of the century when unaccountable narrow lines were observed in the radiation spectrum from the sun. Hypothetical new elements (nebulium and coronium) were invented to explain these lines. It took about three decades until it could be shown that they originate from transitions involving metastable states in highly ionized atoms in the sun's corona."

11. The availability of powerful, high-speed computers makes it possible to analyze data with different cuts in a very short time. This can be crucial in detecting a small signal, but it also has dangers, as we have seen. The ability to vary cuts easily and quickly is shown in Figure 5.14. The graphs show the mass distribution for kaons and pions in the mass region of the D^0 meson. The top row shows the effect of a cut on L/σ, the distance from the primary interaction to the decay vertex divided by the uncertainty in that distance. The prominence of the D^0 peak is enhanced as L/σ gets larger. The bottom row shows those events in which a definite kaon identification has been made. Applying the L/σ cut enhances the peak even more.

12. Not everyone in the community of those who worked on the experiments agreed. Cowan and Greenberg (1996) criticized the APEX result on the grounds that their energy range was too large and that APEX had overestimated the effect that should have been seen in the APEX experiment on the basis of the EPOS I result. APEX disagreed (Ahmad et al., 1996). Greenberg, although a member of the APEX collaboration, withdrew his name from the publication, and has presented a reanalysis of the APEX data that he claims shows peaks similar to those observed earlier. The problem is that

the observation of this effect also requires cuts and one might question whether this result is also an artifact produced by the cuts. Griffin (1995, 1997a,b) has also questioned the conclusions reached by the APEX collaboration and has suggested that there is, in fact, a small peak in their sum-energy spectrum. No criticism of the EPOS II result has been published. Greenberg has requested that APEX continue the search, but the group is not willing to do so. Since the publication of these latest results by EPOS II, APEX, and ORANGE, further evidence against the existence of the sum-energy peaks and the positron lines has been reported by Faestermann and others (1996), Ditzel and others (1997), and Ahmad and collaborators (1997a,b). In particular, the experimenters have investigated the question of whether the peaks observed are due to the internal conversion of the γ rays from nuclear transitions. No effects have been seen.

13. For further discussion of the question of pursuit, see Franklin (1993b).

14. Reference 17 cited by EPOS II in the extract on page 123 is a textbook (Roe, 1992). As the author states, "This book is meant to be a practical introduction into the use of probability and statistics for advanced undergraduate students and for graduate students" (p. v). It includes standard uses of probability in experimental physics and devotes several pages to a discussion of the question of when is a signal significant. The author outlines the method of dividing the data set into two subsets to answer that question:

> In another case, in an international collaboration, we had a group of enthusiasts who had made various cuts and produced a very unexpected $\mu\pi$ resonance in neutrino interactions. We had lots of arguments about whether it was publishable. Fortunately, we were about to analyze the second half of our data. We froze the cuts from the first half and asked whether they produced the same peak in the second half. This is a fair procedure and is useful if you have enough data.
>
> Play with one-half of the data and then if an effect exists, check it in the second half. It is still necessary to derate the probability by the number of times the second half got checked, but it is a considerable help. In our particular case, the signal vanished, but with similar but not quite the same cuts a new signal could be found in μk. (p. 112)

The similarity to the case of the sum-energy peaks is obvious.

This technique of dividing a data set into subsets has also been used to estimate systematic errors in an experiment (Wiss and Gardner, 1994). The point is that it is a standard technique and not unique to EPOS II.

15. It should be emphasized that this data selection cut by Millikan is not unique. Few experiments work properly the first time they are turned on, and no experimenter accepts data unless they are convinced that the apparatus is working properly. In the case of the K_{e2}^{+} branching ratio experiment, I know—because I was a participant in the experiment—that data were excluded when the apparatus was not working properly. I believe that this was also true for the other episodes discussed so far. Consider a problem that developed in the experiment that first demonstrated the violation of combined particle-antiparticle and space-reflection symmetry (CP violation) (Christenson et al., 1964). An interaction with other nearby experiments not only stopped the taking of data, but also led to excluding the data taken while the problem existed. "These runs were interrupted by discovery that bending magnet of Frisch at 6 BeV gives ~20/1 ratio of [counter] 3 to [counter] 2. This is intolerable. Now they have reduced beam and we

resume running pending solution" (quote from the laboratory notebook of the Fitch-Cronin experiment). The setting of a magnet in the adjacent experiment run by an MIT group dramatically changed, and not for the better, the operation of the Princeton experiment. It made the data taken under those conditions unreliable: It wasn't "good" data. For details, see Franklin (1986, pp. 83–87).

16. Consider the discovery of the J/Ψ particle, a particle with an extremely narrow energy width. One of the experiments that originally found the particle used colliding electron and positron beams. Only when each beam had half the energy of the mass of the J/Ψ was the dramatic cross-section increase that signaled the presence of the particle seen. Changing the beam energies slightly caused the phenomenon to disappear. Experimenters thought that this type of effect might be occurring in the heavy-ion collisions that produced the low-mass electron-positron states.

17. The general question of how discordant results are resolved will be discussed in the next section.

6. "Blind" Analysis

1. As we have also seen, the results of Millikan's cosmetic surgery were quite small.

2. A similar argument was used by Robert Millikan to support his observation of the quantization of electric charge and his measurement of the charge of the electron. Millikan remarked, "The total number of changes which we have observed would be between one and two thousand, and *in not one single instance has there been any change which did not represent the advent upon the drop of one definite invariable quantity of electricity or a very small multiple of that quantity*" (Millikan, 1911, p. 360). See the discussion in the Introduction.

3. This is not to say that there was deliberate bias. The selection of data that produced the desired result may have been unconscious.

4. The BABAR group consists of more than 500 physicists. The "Guidelines" were written by the Blind Analysis Task Force and the BABAR Publication Board.

5. For further discussion of this issue, see Franklin (1984; 1986, Chapter 8).

6. This bias is not always fatal: Later in this chapter I discuss a case in which experimental data were analyzed and a result presented. The analysis was modified to include blind analysis and the same experimental data were further analyzed. One could compare the initial results obtained without blind analysis to the later results obtained with such analysis. They differed only slightly.

7. The energy was fixed to produce another particle, the upsilon (4s), which then decayed into the B meson.

8. The $\bar{B}^0$ is the antiparticle of the B^0.

9. The difference is that for the experiments I am discussing, a single offset value was used, whereas in the Fairbank experiment, a different offset value was used for each measurement of the residual charge. In the Fairbank experiment, the offsets used were set by a random number algorithm that was started at a particular number. Only Alvarez knew that number; the offsets were subtracted after the final data were selected.

10. The BABAR task force cited this experiment.

11. The phenomenon of regeneration is a feature of neutral K mesons. The initial beam consists of an equal mixture of K_S and K_L mesons. If the beam line is sufficiently long, as is the case in the KTeV experiment, only the longer-lived K_L mesons will

remain. If the beam then passes through matter, K_S mesons are regenerated. For details, see Franklin (1986, Chapter 3).

12. "To keep the trigger rate at a manageable level, triggers are inhibited by fast veto signals from the regenerator, the MA [mask anticounters], a subset of the photon vetoes, and a downstream hodoscope located behind 4 m of steel to detect muons" (Alavi-Harati et al., 1999, p. 24).

13. Some critics of science, as discussed below, have questioned the use of Monte Carlo simulations in the production of experimental results. This check avoided that objection.

14. The KTeV group did not, in fact, use data collected simultaneously in the analysis of the two decay modes. The $\pi^0\pi^0$ sample used was from data collected in 1996, whereas the $\pi^+\pi^-$ samples were from the first 18 days of data collected in 1997. A software problem in the earlier data had resulted in a large (22%) inefficiency for the $\pi^+\pi^-$ events. The problem was fixed and the later data used. "Finally, using $\pi^+\pi^-$ data from 1996 (collected simultaneously with the $\pi^0\pi^0$ data) instead of from 1997 yields a value of $\text{Re}(\varepsilon'/\varepsilon)$ which is consistent with the standard analysis, allowing a systematic error of 4×10^{-4} due to the 1996 level 3 inefficiency" (Alavi-Harati et al., 1999, p. 27). This was in comparison to the systematic uncertainty of 2.8×10^{-4} in their published result.

15. The branching ratio for $K_L^0 \rightarrow \pi^+\pi^-\pi^0$ is $12.55 \pm 0.20\%$, whereas that for K_{e3}^0 decays is $38.78 \pm 0.28\%$.

16. E is the energy of the particle as measured in the calorimeter and P is the momentum measured in the spectrometer. The π^+ and π^- mesons interact quite differently with the matter in the detector and this gives rise to an asymmetry.

17. There is some question about whether the later analysis could be completely blind. Results, at least for the two decay modes, were already known. The analysis had, however, changed considerably during the intervening three years, and could be considered a new and independent analysis.

18. The D^+ and the D_S^+ are two elementary particles with different masses.

19. Note that the window was chosen independent of the actual data, using a Monte Carlo simulation.

20. In fact, the 1998 "Review of Particle Physics" (Caso et al., 1998) cited only the E791 results for these decays.

21. This was also an important question in the episode of the claimed existence of the 17-keV neutrino, discussed in Chapter 4.

22. "Masked" means excluded from the data set.

23. The lost event was a background event. Had it been a signal event I suspect its loss would have been immediately noticed and a cause for that loss would have been sought.

24. For details of parity nonconservation in the weak interactions, see Franklin (1986, Chapter 1).

25. The information about the blind analysis was presented by Gerry Bunce in a seminar at the University of Colorado and also in private conversation. It does not appear in the published paper.

26. The values (14) and (6) are the statistical and systematic uncertainties in a_μ, respectively.

27. I was in charge of the data analysis for the experiment.

28. A difficulty here is that the endpoint energy of the decay spectra is well known, so that the random offset could also be determined from an examination of the spectra.

29. For another discussion of Monte Carlo calculations, see Galison (1997, Chapter 8).

30. In addition, the input parameters to the Monte Carlo calculations are the best and most reliable values that the experimenters can find.

Part II. The Resolution of Discordant Results

1. The time period needed in the cases I discuss here is of the order of years. Because the resolution of discord often involves the replication of experiments, the construction of new experimental apparatus, and the taking and analyzing of data, this seems to be a reasonable time period.

2. "For all its fallibility, science is the best institution for generating knowledge about the natural world that we have" (Collins 1985, p. 165).

3. In Pickering's later view, in which the stabilization of experimental results is achieved by the mutual adjustments of the theory of the experimental apparatus, the theory of the phenomenon, and the experimental apparatus itself, he omits any discussion of how the discord between experimental results occurs. See my discussion in the Introduction.

7. The Fifth Force

1. This type of calculation, known as upward or downward continuation, was well-known. The results were quite sensitive to the surface-gravity measurements and the model of the earth used. This made knowledge of the local mass distribution and hence the local terrain very important, a point we shall return to later.

2. Typical values for G from mineshaft measurements were $G = (6.720 \pm 0.024) \times 10^{-11}$ m^3 kg^{-1} s^{-2} (Hilton mine) and $6.704^{+0.089}_{-0.025} \times 10^{-11}$ (Mount Isa mine) (Stacey et al., 1987). This should be compared to the best laboratory value at the time of $G = 6.6726(5) \times 10^{-11}$.

3. The Moriond workshops were extremely important in the history of the Fifth Force. At these workshops, many of those working in the field met, presented formal papers, and held informal discussions. If you wanted to be up to date on what was going on in the field, you had to attend these workshops.

4. Other experimental evidence was presented as early as January 1987. The earlier evidence is discussed in the next section.

5. Gravity measurements are generally taken on roads rather than in ditches or surrounding fields. Roads are usually higher than their surroundings, giving rise to an elevation bias.

6. Contrast this with the 16 points in the Greenland survey.

7. Their final result was 60 ± 90 μGal.

8. Parker and Zumberge could not do this for the Australian mine experiments because the data were proprietary.

9. I will not discuss the positive results obtained by Boynton, which were subsequently superseded. This does not change anything essential in the story. For details, see Franklin (1993a).

10. There were, at the time, theoretical explanations that allowed both results to be correct. These were eliminated by further experimental work.

11. Thieberger's experiment was conducted on the Palisades cliff in New Jersey overlooking the Hudson River. His results showed that the float moved away from the cliff. Some wag remarked that all that Thieberger's experiment showed was that any sensible float wanted to leave New Jersey.

8. William Wilson and the Absorption of β Rays

1. The history of this episode is more complex than outlined here. For a time, it appeared that the energy spectrum of electrons from β decay was a line spectrum (i.e., generated by groups of electrons each having the same discrete energy). It was ultimately established that the energy spectrum was continuous, leading to Wolfgang Pauli's suggestion of the neutrino. For details, see Franklin (2000a) and Jensen (2000).

2. The decay products of various elements were sometimes named with a letter or with a numerical suffix, and were later shown to be isotopes of other elements. Thus, radium B was an isotope of lead, ^{214}Pb; radium C was bismuth, ^{214}Bi; and radium E was ^{210}Bi.

3. For details of their work, see Hahn and Meitner (1908a,b, 1909a,b, 1910), von Baeyer and Hahn (1910), and von Baeyer et al. (1911a,b).

4. Wilson's curve was obtained with a radium source, whereas Schmidt had used uranium. Wilson also showed similar results for uranium.

5. For details, see Franklin (2000a, Chapter 1).

6. There was an important experimental problem:

> The field in each electromagnet is a function of the current in both. Thus, if the current through the electromagnet C was kept constant and that in D made to vary, changes took place in both the fields C and D. Now, in the present case, it is required that the field in electromagnet C should be kept constant, while that in D is made to vary, so that changes in the currents in both electromagnets are necessary. The system was therefore calibrated as follows: The current in the electromagnet C was kept constant and that in D varied and the strengths of the fields in each were determined by means of a Grassot fluxmeter, or each value of the current in D. A similar set of readings was taken for about ten different values of the current in C. From the results this obtained, curves could be drawn from which the values of the currents in C and D could be adjusted so that the field in C was kept constant while that in D was made to vary. (Wilson 1910a, pp. 143–44)

7. There is a specific reference to the work of Alois F. Kovarik. Not mentioned are the companion papers on the same subject by Kovarik and Wilson (1910) and Gray and Wilson (1910). The latter paper is discussed in detail below.

8. The line spectra observed were an artifact of the photographic method used to detect the β rays. For details, see Franklin (2000a, pp. 49–50).

9. The Liquid-Scintillator Neutrino Detector

1. For details of this episode, see Franklin (2000a, Chapter 8).

2. Bahcall was the first theorist to calculate the solar neutrino flux and played a leading role in subsequent developments. Davis was the leader of the group that performed the first of the chlorine detector searches for solar neutrinos.

3. The reference is to Davis and collaborators (1968) and Bahcall and collaborators (1968). Davis et al. (1968) is the first experimental report from the Homestake mine chlorine detector. Bahcall et al. (1968) is a theoretical calculation that disagreed with the observation.

4. For this idea to be useful, the neutrino lifetime would have to be less than approximately eight minutes, the time of travel between the sun and the earth—otherwise a significant number would not decay before reaching the earth.

5. A sterile neutrino is one that does not interact with matter.

6. There are actually three types of neutrinos: the electron neutrino, the muon neutrino, and the tau neutrino. For details, see Franklin (2000a, Chapter 7).

7. For details, see Franklin (2000a, Chapter 7); the "Ultimate Neutrino Page," a source for the latest information and results on neutrinos (http://cupp.oulu.fi/neutrino/), compiled by Juha Peltoniemi; and (http://www.sno.phy.queensu.ca/sno/first_results/).

8. The bar denotes an antiparticle. Our best theories, as well as considerable experimental evidence, indicate that every particle has an antiparticle. Particles and antiparticles have the same masses and lifetimes. In the case of charged particles, the antiparticle has the opposite charge of the particle. Thus, the positively charged positron is the antiparticle of the negatively charged electron. Electrically neutral particles such as the neutron and antineutron have different magnetic properties. Other neutral particles, such as the photon (the particle of light) and the neutral pion are their own antiparticles. The question is still open as to whether the neutrinos and antineutrinos are the same or different particles.

9. This was larger, by a factor of four, than the calculated electron-antineutrino component of the beam.

10. For a detailed discussion of selectivity in the production of experimental results, see Part I.

11. The original LSND group used data taken during experimental runs in 1993 and 1994. Hill restricted his analysis to the 1994 data.

12. Notice, once again, the use of "different" experiments to independently confirm a result.

13. Because of the long muon lifetime, a muon that had arrived earlier could decay into an electron and simulate the desired events. Similarly, a muon decay electron could appear later.

14. This is consistent with the oscillation probability obtained in their most recent decay-at-rest experiment, which was $0.31 \pm 0.12 \pm 0.05\%$.

15. For details, see Franklin (2000a, Chapter 9).

16. The SNO group remarked that "The probability that the SNO measurement is not a downward fluctuation from the Super-Kamiokonde measurement is 99.96% (SNO Web site, p. 4)." As of December 2001 the SNO results have not been published.

17. Although the SNO result is independent of the LSND result, the fact that it shows electron-neutrino oscillations gives at least moral support to LSND.

10. Atomic Parity Violation, SLAC E122, and the Weinberg-Salam Theory

1. For a discussion of the discovery of parity nonconservation, see Franklin (1986, Chapter 1).

2. Pickering (1984a) has also discussed this episode from a social constructivist view. Other discussions can be found in Pickering (1991), Ackermann (1991), Lynch (1991), and Franklin (1990, Chapter 8; 1993b).

3. I discuss the uncertainty in the theoretical calculation later.

4. The difference between the two values is $(7.3 \pm 2.5) \times 10^{-8}$, a 2.9 standard-deviation (s.d.) effect, which has a probability 0.37% of being equal to 0 (i.e., an unlikely occurrence). The original experimental result of $(-8 \pm 3) \times 10^{-8}$ cited a two s.d. uncertainty, whereas the later result $(-0.7 \pm 3.2) \times 10^{-8}$ used a 1.5 s.d. uncertainty.

5. Carl Wieman, whose work on atomic parity violation will be discussed below, informed me of this.

6. As we shall see, Dydak's choice was justified by subsequent experimental and theoretical work.

7. Recall that the theoretical predictions of the effect differed by a factor of two.

8. The experimenters used several strategies to establish the validity of their result that I have discussed in this book as parts of an epistemology of experiment. The experimenters intervened and observed the predicted effects when they changed the angle of the calcite prism and when they varied the beam energy. They checked and calibrated their apparatus by using the unpolarized SLAC beam; they observed no instrumental asymmetries and found that their apparatus could measure asymmetries of the expected size. They also used different counters, the lead glass shower counter and the gas Čerenkov counter, and obtained independent confirmation of the validity of their measurement.

9. This result was also presented as an addendum to the Proceedings of the Cargese Workshop.

10. Recall that the Moscow group also saw such effects.

Conclusion

1. It should be emphasized that this data selection cut by Millikan is not unique. Few experiments work properly on their first run, and no experimenter accepts data unless convinced that the apparatus is working properly. In the case of the K^+_{e2} branching ratio experiment, I know—because I was a participant in the experiment—that data were excluded when the apparatus was not working properly. I believe that this was also true for the other episodes discussed in this book. Consider a problem that developed in the experiment that first demonstrated the violation of combined particle-antiparticle and space-reflection symmetry (CP violation) (Christenson et al., 1964). An interaction with other nearby experiments not only stopped the taking of data, but also led to excluding the data taken while the problem existed. "These runs were interrupted by discovery that bending magnet of Frisch at 6 BeV gives ~20/1 ratio of [counter] 3 to [counter] 2. This is intolerable. Now they have reduced beam and we resume running pending solution" (from the laboratory notebook of the Fitch-Cronin experiment). The setting of a magnet in the adjacent experiment run by an MIT group dramatically changed, and not for the better, the operation of the Princeton experiment. It made the data taken under those conditions unreliable: It was not "good" data. For details, see Franklin (1986, pp. 83–87).

2. For a discussion of why different experiments provide more support for a hypothesis than does the repetition of the same experiment, see Franklin and Howson (1984).

3. This is what one might call indirect replication.

4. In this episode, further replications with improved apparatus and critical discussion showed that the original results were incorrect.

5. For more details, see Pickering (1984a, 1991), Collins (1985, 1994), and Franklin (1990, Chapter 8; 1991; 1993b; 1994).

6. Although I suspect that a majority of the physics community was skeptical of the Washington-Oxford results, there were no obvious reasons for believing the results were wrong. The systematic uncertainties that were cited by the Washington and Oxford groups did make the results uncertain.

7. Although both the atomic parity–violation experiments and SLAC E122 used new techniques, they were, in fact, quite different apparatuses, subject to different backgrounds and sources of error and uncertainty. There were no good reasons to lump them together.

8. As noted by Bouchiat (1980) and discussed in detail in Chapter 10, the SLAC E122 experiment was very carefully checked. There were good reasons to believe the result was correct without waiting for an expensive and time-consuming replication. Contrary to Jacqueline Susann, once may be enough.

9. The calculations of the effect predicted by the W-S theory have also changed. Recent calculations have reduced the size of the expected effect.

10. I have never claimed that one must restrict oneself to published sources.

References

Ackermann, R. 1985. *Data, Instruments and Theory*. Princeton, N.J.: Princeton University Press.

———. 1991. "Allan Franklin, Right or Wrong." *PSA 1990, Volume 2*. A. Fine, M. Forbes, and L. Wessels. East Lansing, Mich.: Philosophy of Science Association: 451–57.

Adelberger, E. G. 1989. "High-Sensitivity Hillside Results from the Eot-Wash Experiment." *Tests of Fundamental Laws in Physics: Ninth Moriond Workshop*. O. Fackler and J. Tran Thanh Van. Gif sur Yvette, France: Editions Frontieres: 485–99.

Adelberger, E. G., C. W. Stubbs, W. F. Rogers, et al. 1987. "New Constraints on Composition-Dependent Interactions Weaker than Gravity." *Physical Review Letters* 59: 849–52.

Adler, S., M. S. Atiya, I.-H. Chiang, et al. 1996. "Search for the Decay $K^+ \rightarrow \pi^+ \nu\nu$." *Physical Review Letters* 76: 1421–24.

Ahmad, I., S. M. Austin, B. B. Back, et al. 1995a. "Positron Production in Heavy Ion Collisions: Current Status of the Problem." *Nuclear Physics A* 583: 247–56.

———. 1995b. "Search for Narrow Sum-Energy Lines in Electron-Positron Pair Emission from Heavy-ion Collisions near the Coulomb Barrier." *Physical Review Letters* 75: 2658–61.

———. 1996. "Ahmad et al. Reply." *Physical Review Letters* 77: 2839.

———. 1997a. "Internal Pair Conversion in Heavy Nuclei." *Physical Review C* 55: R2755–59.

———. 1997b. "Search for Monoenergetic Positron Emission from Heavy-Ion Collisions at Coulomb-Barrier Energies." *Physical Review Letters* 78: 618–21.

Aitala, E. M., S. Amato, J. C. Anjos, et al. 1996. "Search for the Flavor-Changing Neutral Current Decays $D^+ \rightarrow \mu^+\mu^-$ and $D^+ \rightarrow e^+e^-$." *Physical Review Letters* 76: 364–67.

———. 1999. "Search for Rare and Forbidden Decays of the D^+, D_s^+, and D^0 Charmed Mesons." *Physics Letters B* 462: 401–9.

Alavi-Harati, A., I. F. Albuquerque, T. Alexopoulos, et al. 1999. "Observation of Direct CP Violation in $K_{S,L} \rightarrow \pi\pi$ Decays." *Physical Review Letters* 83: 22–27.

Altzitzoglou, T., F. Calaprice, M. Dewey, et al. 1985. "Experimental Search for a Heavy Neutrino in the Beta Spectrum of ^{35}S." *Physical Review Letters* 55: 799–802.

Ander, M., M. A. Zumberge, T. Lautzenhiser, et al. 1989. "Test of Newton's Inverse-Square Law in the Greenland Ice Cap." *Physical Review Letters* 62: 985–88.

Anderson, C. D. 1933. "The Positive Electron." *Physical Review* 43: 491–94.

Anselmann, P., W. Hampel, G. Heusser, et al. 1992. "Implications of the GALLEX Determination of the Solar Neutrino Flux." *Physics Letters B* 285: 390–97.

Apalikov, A. M., S. D. Boris, A. I. Golutvin, et al. 1985. "Search for Heavy Neutrinos in β Decay." *JETP Letters* 42: 289–93.

Appel, J. A. 1992. "Hadroproduction of Charm Particles." *Annual Review of Nuclear and Particle Science* 42: 367–99.

Arisaka, K., L. B. Auerbach, S. Axelrod, et al. 1993a. "Improved Sensitivity in a Search for the Rare Decay $K_L^0 \rightarrow e^+e^-$." *Physical Review Letters* 71: 3910–13.

———. 1993b. "Improved Upper Limit on the Branching Ratio $B(K_L^0 \rightarrow \mu^{+-} e^{-+})$." *Physical Review Letters* 70: 1049–52.

Athanassopoulos, C., L. B. Auerbach, D. A. Bauer, et al. 1995. "Candidate Events in a Search for $\nu_\mu \rightarrow \nu_e$ Oscillations." *Physical Review Letters* 75: 2650–53.

Athanassopoulos, C., L. B. Auerbach, R. L. Burman, et al. 1996a. "Evidence for $\nu_\mu \rightarrow \nu_e$ Oscillations from the LSND Experiment at the Los Alamos Meson Physics Facility." *Physical Review Letters* 77: 3082–85.

———. 1996b. "Evidence for Neutrino Oscillations from Muon Decay at Rest." *Physical Review C* 54: 2685–2708.

Athanassopoulos, C., L. B. Auerbach, D. A. Bauer, et al. 1997. "The Liquid Scintillator Neutrino Detector and LAMPF Neutrino Source." *Nuclear Instruments and Methods in Physics Research A* 388: 149–72.

———. 1998a. "Results on $\nu_\mu \rightarrow \nu_e$ Neutrino Oscillations from the LSND Experiment." *Physical Review Letters* 81: 1774–77.

———. 1998b. "Results on $\nu_\mu \rightarrow \nu_e$ Oscillations from Pion Decay in Flight Neutrinos." *Physical Review C* 58: 2489–2511.

Backe, H., L. Handschug, F. Hessberger, et al. 1978. "Observation of Positron Creation In Superheavy Ion-Atom Collision Systems." *Physical Review Letters* 40: 1443–46.

Backe, H., W. Bonin, E. Kankeleit, et al. 1983a. "Positrons from Heavy Ion Collisions—Experiments at the UNILAC." *Quantum Electrodynamics of Strong Fields*. W. Greiner. New York: Plenum Press: 107–32.

Backe, H., P. Senger, W. Bonin, et al. 1983b. "Estimates of the Nuclear Time Delay in Dissipative U + U and U + Cm Collisions Derived from the Shape of Positron and δ-Ray Spectra." *Physical Review Letters* 50: 1838–41.

Bahcall, J. N., N. A. Bahcall, and G. Shaviv. 1968. "Present Status of the Theoretical Predictions for the ^{37}Cl Solar-Neutrino Experiment." *Physical Review Letters* 20: 1209–12.

Bahcall, J. N., and R. Davis. 1989. "An Account of the Development of the Solar Neutrino Problem." *Neutrino Astrophysics*. J. N. Bahcall. Cambridge: Cambridge University Press: 487–530.

Baird, P. E. G., M. W. S. Brimicombe, G. J. Roberts, et al. 1976. "Search for Parity Non-Conserving Optical Rotation in Atomic Bismuth." *Nature* 264: 528–29.

Baird, P. E. G., M. W. S. Brimicombe, R. G. Hunt, et al. 1977. "Search for Parity-Nonconserving Optical Rotation in Atomic Bismuth." *Physical Review Letters* 39: 798–801.

Bar, R., A. Balanda, J. Baumann, et al. 1995. "Experiments on e^+e^--Line Emission in HI Collisions." *Nuclear Physics A* 583: 237–46.

Bargholtz, C., L. Holmberg, K. E. Johansson, et al. 1987. "Investigation of Anomalous Spectral Structure in Low-Energy Positron Scattering." *Journal of Physics G* 13: L265–70.

Barkov, L. M., and M. S. Zolotorev. 1978a. "Observations of Parity Nonconservation in Atomic Transitions." *JETP Letters* 27: 357–61.

———. 1978b. "Measurement of Optical Activity of Bismuth Vapor." *JETP Letters* 28: 503–6.

———. 1979a. "Parity Violation in Atomic Bismuth." *Physics Letters B* 85: 308–13.

———. 1979b. "Parity Violation in Bismuth: Experiment." *International Workshop on Neutral Current Interactions in Atoms in Cargese.* W. L. Williams. Washington: National Science Foundation: 52–76.

———. 1980. "Parity Nonconservation in Bismuth Atoms and Neutral Weak-Interaction Currents." *JETP* 52: 360–69.

Barnes, B. 1991. "How Not to Do the Sociology of Knowledge." *Rethinking Objectivity, Part 1. Special Issue of Annals of Scholarship* 8: 321–35.

Bartlett, D. F., and W. L. Tew. 1989a. "The Fifth Force: Terrain and Pseudoterrain." *Tests of Fundamental Laws in Physics: Ninth Moriond Workshop.* O. Fackler and J. Tran Thanh Van. Gif sur Yvette, France: Editions Frontieres: 543–48.

———. 1989b. "Possible Effect of the Local Terrain on the Australian Fifth-Force Measurement." *Physical Review D* 40: 673–75.

Becquerel, H. 1900. "Sur la transparence de l'aluminium pour la rayonnement du radium." *Comptes Rendus des Seances de L'Academie des Sciences* 130: 1154–57.

Belz, J., R. D. Cousins, M. V. Diwan, et al. 1996. "Search for the Weak Decay of an H Dibaryon." *Physical Review Letters* 76: 3277–80.

Benvenuti, A., D. Cline, F. Messing, et al. 1976. "Evidence for Parity Nonconservation in the Weak Neutral Current." *Physical Review Letters* 37: 1039–42.

Birich, G. N., Y. V. Bogdanov, S. I. Kanorskii, et al. 1984. "Nonconservation of Parity in Atomic Bismuth." *JETP* 60: 442–49.

Bizzeti, P. G., A. M. Bizzeti-Sona, T. Fazzini, et al. 1988. "New Search for the 'Fifth Force' with the Floating Body Method: Status of the Vallambrosa Experiment." *Fifth Force Neutrino Physics: Eighth Moriond Workshop.* O. Fackler and J. Tran Thanh Van. Gif sur Yvette, France: Editions Frontieres: 501–13.

———. 1989. "Search for a Composition-Dependent Fifth Force." *Physical Review Letters* 62: 2901–4.

Bogdanov, Y. V., I. I. Sobel'man, V. N. Sorokin, et al. 1980a. "Investigation of Optical Activity of Bi Vapors." *JETP Letters* 31: 214–19.

———. 1980b. "Parity Nonconservation in Atomic Bismuth." *JETP Letters* 31: 522–26.

Bokemeyer, H., K. Bethge, H. Folger, et al. 1983. "Search for Spontaneous Positron Production in Heavy-Ion Collisions." *Quantum Electrodynamics of Strong Fields*. W. Greiner. New York: Plenum Press: 273–92.

Bokemeyer, H., P. Salabura, D. Schwalm, et al. 1989. "Correlated Electron-Positron Emission in Heavy-Ion Collisions." *Tests of Fundamental Laws in Physics*. O. Fackler and J. Tran Thanh Van. Gif sur Yvette, France: Editions Frontieres: 77–92.

Bonvicini, G. 1993. "Statistical Issues in the 17-keV Neutrino Experiments." *Zeitschrift fur Physik A* 345: 97–117.

Borge, M. J. G., A. De Rujula, P. G. Hansen, et al. 1986. "Limits on Neutrino-Mixing from the Internal Bremsstrahlung Spectrum of ^{125}I." *Physica Scripta* 34: 591–96.

Bouchiat, C. 1980. "Neutral Current Interactions in Atoms." *International Workshop on Neutral Current Interactions in Atoms*. W. L. Williams. Washington: National Science Foundation: 357–69.

Bouchiat, M. A., and C. Bouchiat. 1974. "Weak Neutral Currents in Atomic Physics." *Physics Letters B* 48: 111–14.

Bouchiat, M. A., and L. Pottier. 1984. "Atomic Parity Violation Experiments." *Atomic Physics 9*. R. Van Dyck and E. Fortson. Singapore: World Scientific: 246–71.

Bouchiat, M. A., J. Guena, L. Hunter, et al. 1982. "Observation of Parity Violation in Cesium." *Physics Letters B* 117: 358–64.

Bouchiat, M. A., J. Guena, and L. Pottier. 1984. "New Observation of a Parity Violation in Cesium." *Physics Letters B* 134: 463–68.

Bowen, D. R., A. K. Mann, W. K. McFarlane, et al. 1967. "Measurement of the K^{+}_{e2} Branching Ratio." *Physical Review* 154: 1314–22.

Bragg, W. H. 1904. "On the Absorption of Alpha Rays and on the Classification of Alpha Rays from Radium." *Philosophical Magazine* 8: 719–25.

Brown, H. N., G. Bunce, R. M. Carey, et al. 2001. "Precise Measurement of the Positive Muon Anomalous Magnetic Moment." *Physical Review Letters* 86: 2227–31.

Bucksbaum, P., E. Commins, and L. Hunter. 1981a. "New Observation of Parity Nonconservation in Atomic Thallium." *Physical Review Letters* 46: 640–43.

———. 1981b. "Observation of Parity Nonconservation in Atomic Thallium." *Physical Review* 24D: 1134–38.

Burchat, P., T. Champion, B. Ford, et al. 2000. Draft Guidelines for Blind Analyses in BABAR. Palo Alto, Calif.: SLAC.

Cabrera, B. 1982. "First Results from a Superconductive Detector for Moving Magnetic Monopoles." *Physical Review Letters* 48: 1378–81.

Caso, C., G. Conforto, A. Gurtu, et al. 1998. "Review of Particle Physics." *European Physical Journal* 3: 1–794.

Christenson, J. H., J. W. Cronin, V. L. Fitch, et al. 1964. "Evidence for the 2π Decay of the K^{0}_{2} Meson." *Physical Review Letters* 13: 138–40.

Clemente, M., E. Berdermann, P. Kienle, et al. 1984. "Narrow Positron Lines from U-U and U-Th Collisions." *Physics Letters B* 137: 41–46.

Close, F. E. 1976. "Parity Violation in Atoms?" *Nature* 264: 505–6.

Collins, H. 1981. "Stages in the Empirical Programme of Relativism." *Social Studies of Science* 11: 3–10.

———. 1985. *Changing Order: Replication and Induction in Scientific Practice*. London: Sage Publications.

————. 1994. "A Strong Confirmation of the Experimenters' Regress." *Studies in History and Philosophy of Modern Physics* 25: 493–503.

Collins, H., and T. Pinch. 1993. *The Golem: What Everyone Should Know About Science.* Cambridge: Cambridge University Press.

Conan Doyle, A. 1967. "The Sign of Four." *The Annotated Sherlock Holmes.* W. S. Barrington-Gould. New York: Clarkson N. Potter.

Connell, S. H., R. W. Fearick, R. F. A. Hoernle, et al. 1988. "Search for Low-Energy Resonances in the Electron-Positron Annihilation-in-Flight Photon Spectrum." *Physical Review Letters* 60: 2242–45.

Conti, R., P. Bucksbaum, S. Chu, et al. 1979. "Preliminary Observation of Parity Nonconservation in Atomic Thallium." *Physical Review Letters* 42: 343–46.

Cowan, T. E., and J. S. Greenberg. 1996. "Comment on the APEX e^+e^- Experiment." *Physical Review Letters* 77: 2839.

Cowan, T., H. Backe, M. Begemann, et al. 1985. "Anomalous Positron Peaks from Supercritical Collision Systems." *Physical Review Letters* 54: 1761–64.

Cowan, T., H. Backe, K. Bethge, et al. 1986. "Observation of Correlated Narrow-Peak Structures in Positron and Electron Spectra from Superheavy Collision Systems." *Physical Review Letters* 56: 444–47.

Cowan, T. E., J. S. Greenberg, H. Backe, et al. 1987. "Narrow Correlated Electron-Positron Peaks from Superheavy Collision Systems." *Physics of Strong Fields.* W. Greiner. New York: Plenum Press: 111–94.

Creath, R. 1995. "Are Dinosaurs Extinct?" *Foundations of Science* 1: 285–97.

Crowther, J. A. 1910. "On the Transmission of β-rays." *Proceedings of the Cambridge Philosophical Society* 15: 442–58.

Dancoff, S. M. 1952. "Does the Neutrino *Really* Exist?" *Bulletin of the Atomic Scientists* 8: 139–41.

Datar, V. M., C. V. K. Baba, S. K. Bhattacherjee, et al. 1985. "Search for a Heavy Neutrino in the β-decay of ^{35}S." *Nature* 318: 547–48.

Davis, R., D. S. Harmer, and K. C. Hoffman. 1968. "Search for Neutrinos from the Sun." *Physical Review Letters* 20: 1205–9.

Ditzel, E., J. M. Hoogduin, H. Backe, et al. 1997. "Absence of E0 Transitions around 1.8 MeV after Collisions of ^{238}U with ^{181}Ta." *Zeitschrift fur Physik A* 358: 11–14.

Douglass, D. H., R. Q. Gram, J. A. Tyson, et al. 1975. "Two-Detector-Coincidence Search for Bursts of Gravitational Radiation." *Physical Review Letters* 35: 480–83.

Dydak, F. 1979. "Neutral Currents." *International Conference on High Energy Physics.* Geneva: CERN: 25–49.

Drukarev, E. G., and M. I. Strikman. 1986. "Final-state Interaction of β Electrons and Related Phenomena." *JETP* 64: 686–92.

Eckhardt, D. 1989. "Evidence for Non-Newtonian Gravity: Status of the AFGL Experiment January 1989." *Tests of Fundamental Laws in Physics: Ninth Moriond Workshop.* O. Fackler and J. Tran Thanh Van. Gif sur Yvette, France: Editions Frontieres: 525–27.

Eckhardt, D. H., C. Jekeli, A. R. Lazarewicz, et al. 1988. "Results of a Tower Gravity Experiment." *Fifth Force Neutrino Physics: Eighth Moriond Workshop.* O. Fackler and J. Tran Thanh Van. Gif sur Yvette, France: Editions Frontieres: 577–83.

Ellis, C. D., and W. A. Wooster. 1927. "The Average Energy of Disintegration of Radium E." *Proceedings of the Royal Society (London)* A117: 109–23.

Eman, B., and D. Tadic. 1986. "Distortion in the β-decay Spectrum for Low Electron Kinetic Energies." *Physical Review C* 33: 2128–31.

Emmons, T. P., J. M. Reeve, and E. N. Fortson. 1983. "Parity-Nonconserving Optical Rotation in Atomic Lead." *Physical Review Letters* 53: 2089–92.

Eötvös, R., D. Pekar, and E. Fekete. 1922. "Beitrage zum Gesetze der Proportionalitat von Tragheit und Gravitat." *Annalen der Physik (Leipzig)* 68: 11–66.

Erb, K. A., I. Y. Lee, and W. T. Milner. 1986. "Evidence for Peak Structure in $e^+ + $ Th Interactions." *Physics Letters B* 181: 52–56.

Faestermann, T., F. Heine, and P. Kienle. 1996. "Are the GSI Positron Electron Lines Due to Internally Converted E0 Transitions?" *Zeitschrift fur Physik A* 354: 213–14.

Fairbank, W. M. 1988. "Summary Talk on Fifth Force Papers." *5th Force Neutrino Physics: Eighth Moriond Workshop.* O. Fackler and J. Tran Thanh Van. Gif sur Yvette: Editions Frontieres: 629–44.

Fairbank Jr., W. M., and A. Franklin. 1982. "Did Millikan Observe Fractional Charges on Oil Drops?" *American Journal of Physics* 50: 394–97.

Feynman, R. P., R. B. Leighton, and M. Sands. 1963. *The Feynman Lectures on Physics.* Reading, Mass.: Addison-Wesley.

Fischbach, E., S. Aronson, C. Talmadge, et al. 1986. "Reanalysis of the Eötvös Experiment." *Physical Review Letters* 56: 3–6.

Fortson, E. N., and L. L. Lewis. 1984. "Atomic Parity Nonconservation." *Physics Reports* 113: 289–344.

Franklin, A. 1981. "Millikan's Published and Unpublished Data on Oil Drops." *Historical Studies in the Physical Sciences* 11: 185–201.

———. 1984. "Forging, Cooking, Trimming, and Riding on the Bandwagon." *American Journal of Physics* 52: 786–93.

———. 1986. *The Neglect of Experiment.* Cambridge: Cambridge University Press.

———. 1988. "Experiment, Theory Choice, and the Duhem-Quine Problem." *Theory and Experiment.* D. Batens and J. P. van Bendegem. Dordrecht: Reidel: 141–55.

———. 1990. *Experiment, Right or Wrong.* Cambridge: Cambridge University Press.

———. 1991. "Do Mutants Have to Be Slain, or Do They Die of Natural Causes?" *PSA 1990, Volume 2.* A. Fine, M. Forbes and L. Wessels. East Lansing, Mich.: Philosophy of Science Association: 487–94.

———. 1993a. *The Rise and Fall of the Fifth Force: Discovery, Pursuit, and Justification in Modern Physics.* New York: American Institute of Physics.

———. 1993b. "Discovery, Pursuit, and Justification." *Perspectives on Science* 1: 252–84.

———. 1993c. "Experimental Questions." *Perspectives on Science* 1: 127–46.

———. 1994. "How to Avoid the Experimenters' Regress." *Studies in History and Philosophy of Modern Physics* 25: 97–121.

———. 1995a. "The Appearance and Disappearance of the 17-keV Neutrino." *Reviews of Modern Physics* 67: 457–90.

———. 1995b. "Laws and Experiment." *Laws of Nature.* F. Weinert. Berlin: De Gruyter: 191–207.

———. 1996. "There Are No Antirealists in the Laboratory." *Realism and Anti-Realism in the Philosophy of Science.* R. S. Cohen, R. Hilpinen, and Q. Renzong. Dordrecht: Kluwer Academic: 131–48.

———. 1997a. "Calibration." *Perspectives on Science* 5: 31–80.

———. 1997b. "Are There Really Electrons? Experiment and Reality." *Physics Today* 50: 26–33.

———. 1998. "Selectivity and the Production of Experimental Results." *Archive for the History of Exact Sciences* 53: 399–485.

———. 1999. *Can That Be Right? Essays on Experiment, Evidence, and Science.* Dordrecht: Kluwer Academic.

———. 2000a. *Are There Really Neutrinos? An Evidential History.* Cambridge, Mass.: Perseus Books.

———. 2000b. "The Road to the Neutrino." *Physics Today* 53: 22–28.

———. 2002. "William Wilson and the Absorption of Beta Rays." *Physics in Perspective* 4: 50–70.

Franklin, A., and C. Howson. 1984. "Why Do Scientists Prefer to Vary Their Experiments?" *Studies in History and Philosophy of Science* 15: 51–62.

———. 1988. "It Probably is a Valid Experimental Result: A Bayesian Approach to the Epistemology of Experiment." *Studies in the History and Philosophy of Science* 19: 419–27.

———. 1998. "Comment on 'The Structure of a Scientific Paper' by Frederick Suppe." *Philosophy of Science* 65: 411–16.

Galison, P. 1987. *How Experiments End.* Chicago: University of Chicago Press.

———. 1997. *Image and Logic: A Material Culture of Microphysics.* Chicago: University of Chicago Press.

———. 1999. "Reflections on *Image and Logic: A Material Culture of Microphysics.*" *Perspectives on Science* 7: 255–84.

Ganz, R., R. Bar, A. Balanda, et al. 1996. "Search for e^+e^- Pairs with Narrow Sum-Energy Distributions in Heavy Ion Collisions." *Physics Letters B* 389: 4–12.

Garwin, R. L. 1974. "Detection of Gravity Waves Challenged." *Physics Today* 27: 9–11.

Gell-Mann, M. 1964. "A Schematic Model of Baryons and Mesons." *Physics Letters* 8: 214–15.

Gilbert, S. L., and C. E. Wieman. 1986. "Atomic Beam Measurement of Parity Nonconservation in Cesium." *Physical Review* 34A: 792–803.

Gilbert, S. L., M. C. Noecker, R. N. Watts, et al. 1985. "Measurement of Parity Nonconservation in Atomic Cesium." *Physical Review Letters* 55: 2680–83.

Glashow, S. L. 1991. "A Novel Neutrino Mass Hierarchy." *Physics Letters B* 256: 255–57.

———. 1992. "The Death of Science!?" *The End of Science? Attack and Defense.* R. J. Elvee. Lanham, Md.: University Press of America.

Gooding, D. 1992. "Putting Agency Back into Experiment." *Science as Practice and Culture.* A. Pickering. Chicago: University of Chicago Press: 65–112.

Gray, J. A. 1910. "The Distribution of Velocity in the β-Rays from a Radioactive Substance." *Proceedings of the Royal Society (London)* A84: 136–41.

Gray, J. A., and W. Wilson. 1910. "The Heterogeneity of the β Rays from a Thick Layer of Radium E." *Philosophical Magazine* 20: 870–75.

Greenberg, J. S. 1983. "Summary: Experimental Aspects." *Quantum Electrodynamics of Strong Fields.* W. Greiner. New York: Plenum Press: 853–92.

Greenberg, J. S., and W. Greiner. 1982. "Search for the Sparking of the Vacuum." *Physics Today* 35: 24–32.

Greiner, W., and J. Reinhardt. 1995. "Quo Vadis Electron-Positron Lines?" *Physica Scripta* T56: 203–19.

Griffin, J. J. 1995. "Sharp (e^+e^-) Pairs: Reason for Caution." *Acta Physica Slovaca* 45: 661–72.

———. 1997a. Comment on "A Search for Narrow Sum Energy Lines in Electron-Positron Pair Emission from Heavy Ion Collisions Near the Coulomb Barrier." College Park, Md.: University of Maryland.

———. 1997b. APEX: Positive Evidence for Sharp 800keV Pairs from Heavy Ion Collisions Near the Coulomb Barrier. College Park, Md: University of Maryland.

Groom, D., M. Aguilar-Benitez, C. Amsler, et al. 2000. "Review of Particle Physics." *European Physical Journal* 15: 1–878.

Hacking, I. 1981. "Do We See Through a Microscope?" *Pacific Philosophical Quarterly* 63: 305–22.

———. 1983. *Representing and Intervening*. Cambridge: Cambridge University Press.

———. 1992. "The Self-Vindication of the Laboratory Sciences." *Science as Practice and Culture*. A. Pickering. Chicago: University of Chicago Press: 29–64.

———. 1999. *The Social Construction of What?* Cambridge, Mass.: Harvard University Press.

Hahn, O. 1966. *Otto Hahn: A Scientific Autobiography*. New York: Charles Scribner's Sons.

Hahn, O., and L. Meitner. 1908a. "Uber die Absorption der β-Strahlen einiger Radioelemente." *Physikalische Zeitschrift* 9: 321–33.

———. 1908b. "Uber die β -Strahlen des Aktiniums." *Physikalische Zeitschrift* 9: 697–704.

———. 1909a. "Uber eine typische β -Strahlung des eigentlicher Radiums." *Physikalische Zeitschrift* 10: 741–45.

———. 1909b. "Uber das Absrorptionsgesetz der β -Strahlen." *Physikalische Zeitschrift* 10: 948–50.

———. 1910. "Eine neue β -Strahlung bein Thorium X; Analogein in der Uran- und Thoriumreihe." *Physikalische Zeitschrift* 11: 493–97.

Harding, S., ed. 1976. *Can Theories Be Refuted?* Dordrecht: Reidel.

———. 1996. "Science is 'Good to Think With.'" *Social Text* 46–47: 15–26.

Haxton, W. C. 1985. "Atomic Effects and Heavy Neutrino Emission in Beta Decay." *Physical Review Letters* 55: 807–9.

Heilbron, J. L. 1967. "The Scattering of α and β Particles and Rutherford's Atom." *Archive for History of Exact Sciences* 4: 247–307.

Hetherington, D. W., R. L. Graham, M. A. Lone, et al. 1986. "Search for Evidence of a 17-keV Neutrino in the Beta Spectrum of ^{63}Ni." *Nuclear Beta Decays and Neutrino: Proceedings of the International Symposium, Osaka, Japan, June 1986*. T. Kotani, H. Ejiri, and E. Takasugi. Singapore: World Scientific: 387–90.

———. 1987. "Upper Limits on the Mixing of Heavy Neutrinos in the Beta Decay of ^{63}Ni." *Physical Review C* 36: 1504–13.

Hill, J. E. 1995. "An Alternative Analysis of the LSND Neutrino Oscillation Search Data on $\nu_\mu \rightarrow \nu_e$." *Physical Review Letters* 75: 2654–57.

Hime, A. 1992. "Pursuing the 17 keV Neutrino." *Modern Physics Letters A* 7: 1301–14.

Hime, A. 1993. "Do Scattering Effects Resolve the 17-keV Conundrum?" *Physics Letters B* 299: 165–73.

Hime, A., and N. A. Jelley. 1991. "New Evidence for the 17 keV Neutrino." *Physics Letters B* 257: 441–49.

Hime, A., and J. J. Simpson. 1989. "Evidence of the 17-keV Neutrino in the β Spectrum of ^{3}H." *Physical Review D* 39: 1837–50.

Hirata, K. S., T. Kajita, M. Koshiba, et al. 1988. "Observation in the Kamiokonde-II Detector of the Neutrino Burst from Supernova SN1987A." *Physical Review D* 38: 448–58.

Hollister, J. H., G. R. Apperson, L. L. Lewis, et al. 1981. "Measurement of Parity Nonconservation in Atomic Bismuth." *Physical Review Letters* 46: 643–46.

Howson, C., and A. Franklin. 1986. "A Bayesian Analysis of Excess Content and the Localisation of Support." *British Journal for the Philosophy of Science* 36: 425–31.

Imlay, R. L., P. T. Eschstruth, A. D. Franklin, et al. 1967. "Energy Dependence of the Form Factor in K^+_{e3} Decay." *Physical Review* 160: 1203–11.

Jekeli, C., D. H. Eckhardt, and A. J. Romaides. 1990. "Tower Gravity Experiment: No Evidence for Non-Newtonian Gravity." *Physical Review Letters* 64: 1204–6.

Jensen, C. 2000. *Controversy and Consensus: Nuclear Beta Decay 1911–1934*. Boston: Birkhauser.

Jones, L. W. 1977. "A Review of Quark Search Experiments." *Reviews of Modern Physics* 49: 717–52.

Kalbfleisch, G. R., and K. A. Milton. 1985. "Heavy-Neutrino Emission." *Physical Review Letters* 55: 2225.

Kammeraad, J., P. Kasameyer, O. Fackler, et al. 1990. "New Results from Nevada: A Test of Newton's Law Using the BREN Tower and a High Density Gravity Survey." *New and Exotic Phenomena '90: Tenth Moriond Workshop*. O. Fackler and J. Tran Thanh Van. Gif sur Yvette, France: Editions Frontieres: 245–54.

Kasameyer, P., J. Thomas, O. Fackler, et al. 1989. "A Test of Newton's Law of Gravity Using the BREN Tower, Nevada." *Tests of Fundamental Laws in Physics: Ninth Moriond Workshop*. O. Fackler and J. Tran Thanh Van. Gif sur Yvette, France: Editions Frontieres: 529–42.

Kaufmann, W. 1902. "Die elektromagnetische Masse des Elektrons." *Physikalische Zeitschrift* 4: 54–57.

Kawakami, H., S. Kato, T. Ohshima, et al. 1992. "High Sensitivity Search for a 17 keV Neutrino. Negative Indication with an Upper Limit of 0.095%." *Physics Letters B* 287: 45–50.

Kienle, P. 1983. "Pair Creation in Heavy Ion-Atom Collisions." *Quantum Electrodynamics of Strong Fields*. W. Greiner. New York: Plenum Press: 293–316.

———. 1989. "Electron-Positron Pair Creation in Heavy Ion Collisions Investigated with 'Orange' Spectrometers." *Tests of Fundamental Laws in Physics*. O. Fackler and J. Tran Thanh Van. Gif sur Yvette: Editions Frontieres: 63–75.

Koenig, W., F. Bosch, P. Kienle, et al. 1987. "Positron Lines from Subcritical Heavy Ion-Atom Collisions." *Zeitschrift fur Physik A* 328: 129–45.

Koenig, W., E. Berdermann, F. Bosch, et al. 1989. "On the Momentum Correlation of (e^+e^-) Pairs Observed in U + U and U + Pb Collisions." *Physics Letters B* 218: 12–16.

————. 1993. "Investigations of Correlated e^+e^- Emission in Heavy-Ion Collisions near the Coulomb Barrier." *Zeitschrift fur Physik A* 346: 153–68.

Koertge, N., ed. 1998. *A House Built on Sand: Exposing Postmodernist Myths About Science*. Oxford: Oxford University Press.

Kovarik, A. F., and W. Wilson. 1910. "On the Reflexion of Homogeneous β-Particles of Different Velocities." *Philosophical Magazine* 20: 866–70.

Kozhuharov, C., P. Kienle, E. Berdermann, et al. 1979. "Positrons from 1.4-GeV Uranium Atom Collisions." *Physical Review Letters* 42: 376–79.

Kuroda, K., and N. Mio. 1989. "Test of a Composition-Dependent Force by a Free-Fall Interferometer." *Physical Review Letters* 62: 1941–44.

LaRue, G. S., J. D. Phillips, and W. M. Fairbank. 1981. "Observation of Fractional Charge of $(1/3)e$ on Matter." *Physical Review Letters* 46: 967–70.

Latour, B., and S. Woolgar. 1979. *Laboratory Life: The Social Construction of Scientific Facts*. Beverly Hills: Sage.

————. 1986. *Laboratory Life: The Construction of Scientific Facts*. Princeton: Princeton University Press.

Leinberger, U., E. Berdermann, F. Heine, et al. 1997. "New Results on e^+e^--line Emission in U + Ta Collisions." *Physics Letters B* 394: 16–22.

Levine, J. L., and R. L. Garwin. 1974. "New Negative Result for Gravitational Wave Detection, and Comparison with Reported Detection." *Physical Review Letters* 33: 794–97.

Lewis, L. L., J. H. Hollister, D. C. Soreide, et al. 1977. "Upper Limit on Parity-Nonconserving Optical Rotation in Atomic Bismuth." *Physical Review Letters* 39: 795–98.

Lindhard, J., and P. G. Hansen. 1986. "Atomic Effects in Low-Energy Beta Decay: The Case of Tritium." *Physical Review Letters* 57: 965–67.

Lipton, P. 1998. "The Best Explanation of a Scientific Paper." *Philosophy of Science* 65: 406–10.

Lorenz, E., G. Mageras, U. Stiegler, et al. 1988. "Search for Narrow Resonance Productions in Bhabha Scattering at Center-of-Mass Energies Near 1.8 MeV." *Physics Letters B* 214: 10–13.

Louis, B., V. Sandberg, H. White, et al. 1997. "A Thousand Eyes: The Story of LSND." *Los Alamos Science* 25: 92–115.

Lynch, M. 1991. "Allan Franklin's Transcendental Physics." *PSA 1990, Volume 2*. A. Fine, M. Forbes, and L. Wessels. East Lansing, Mich.: Philosophy of Science Association: 471–85.

MacKenzie, D. 1989. "From Kwajelein to Armageddon? Testing and the Social Construction of Missile Accuracy." *The Uses of Experiment*. D. Gooding, T. Pinch, and S. Shaffer. Cambridge: Cambridge University Press: 409–35.

Maier, K., W. Bauer, J. Briggmann, et al. 1987. "Experimental Limits for Narrow Lines in the Excitation Function of Positron-Electron Scattering around $E^* = 620$ keV and $E^* = 810$ keV." *Zeitschrift fur Physik A* 326: 527–29.

Maier, K., E. Widmann, W. Bauer, et al. 1988. "Evidence for a Resonance in Positron-Electron Scattering at 810 keV Centre-of-Mass Energy." *Zeitschrift fur Physik A* 330: 173–81.

Markey, H., and F. Boehm. 1985. "Search for Admixture of Heavy Neutrinos with Masses between 5 and 55 keV." *Physical Review C* 32: 2215–16.

Meitner, L. 1964. "Looking Back." *Bulletin of the Atomic Scientists* 20: 2–7.

Meyer, S., and E. von Schweidler. 1899. "Uber das Verhalten von Radium und Polonium im Magnetischen Felde (I. Mitteilung)." *Physikalische Zeitschrift* 1: 90–91.

Miller, D. J. 1977. "Elementary Particles—A Rich Harvest." *Nature* 269: 286–88.

Millikan, R. A. 1911. "The Isolation of an Ion, A Precision Measurement of Its Charge, and the Correction of Stokes's Law." *Physical Review* 32: 349–97.

———. 1913. "On the Elementary Electrical Charge and the Avogadro Constant." *Physical Review* 2: 109–43.

———. 1917. *The Electron.* Chicago: University of Chicago Press.

Mills Jr., A. P., and J. Levy. 1987. "Search for a Bhabha-scattering Resonance near 1.8 MeV/c^2." *Physical Review D* 36: 707–12.

Morrison, D. 1992. "Review of 17 keV Neutrino Experiments." *Joint International Lepton-Photon Symposium and Europhysics Conference on High Energy Physics.* S. Hegarty, K. Potter, and E. Quercigh. Geneva, Switzerland: World Scientific. 1: 599–605.

Mortara, J. L., I. Ahmad, K. P. Coulter, et al. 1993. "Evidence Against a 17 keV Neutrino from ^{35}S Beta Decay." *Physical Review Letters* 70: 394–97.

Nelson, A. 1994. "How Could Scientific Facts be Socially Constructed?" *Studies in History and Philosophy of Science* 25: 535–47.

Nguyen, H. 2001. Long Writeup for the Measurement of δ_L in 1997 KTeV Data, KTeV Group (Internal report).

Niebauer, T. M., M. P. McHugh, and J. E. Faller. 1987. "Galilean Test for the Fifth Force." *Physical Review Letters* 59: 609–12.

Ohi, T., M. Nakajima, H. Tamura, et al. 1985. "Search for Heavy Neutrinos in the Beta Decay of ^{35}S. Evidence against the 17 keV Heavy Neutrino." *Physics Letters B* 160: 322–24.

Ohshima, T. 1993. "0.073% (95% CL) Upper Limit on 17 keV Neutrino Admixture." *XXVI International Conference on High Energy Physics.* J. R. Sanford. Dallas: American Institute of Physics. 1: 1128–35.

Ohshima, T., H. Sakamoto, T. Sato, et al. 1993. "No 17 keV Neutrino: Admixture < 0.073% (95% C.L.)." *Physical Review D* 47: 4840–56.

Pais, A. 1986. *Inward Bound.* New York: Oxford University Press.

Parker, R. L., and M. A. Zumberge. 1989. "An Analysis of Geophysical Experiments to Test Newton's Law of Gravity." *Nature* 342: 29–32.

Passmore, J. 1972. "Logical Positivism." *Encyclopedia of Philosophy.* P. Edwards. New York: Collier-MacMillan: 52–57.

Peckhaus, R., T. W. Elze, T. Happ, et al. 1987. "Search for Peak Structure in e^+ + Th Collisions." *Physical Review C* 36: 83–86.

Pevsner, A., R. Kraemer, M. Nussbaum, et al. 1961. "Evidence for a Three-Pion Resonance near 550 MeV." *Physical Review Letters* 4: 412–23.

Phillips, J. D. (1980). Residual Charge of Niobium Spheres. *Physics.* Ph.D. dissertation, Stanford University.

Pickering, A. 1981. "The Hunting of the Quark." *Isis* 72: 216–36.

————. 1984a. *Constructing Quarks*. Chicago: University of Chicago Press.

————. 1984b. "Against Putting the Phenomena First: The Discovery of the Weak Neutral Current." *Studies in the History and Philosophy of Science* 15: 85–117.

————. 1987. "Against Correspondence: A Constructivist View of Experiment and the Real." *PSA 1986, Volume 2*. A. Fine and P. Machamer. Pittsburgh: Philosophy of Science Association: 196–206.

————. 1989. "Living in the Material World: On Realism and Experimental Practice." *The Uses of Experiment*. D. Gooding, T. Pinch, and S. Schaffer. Cambridge: Cambridge University Press: 275– 97.

————. 1991. "Reason Enough? More on Parity Violation Experiments and Electroweak Gauge Theory." *PSA 1990, Volume 2*. A. Fine, M. Forbes, and L. Wessels. East Lansing, Mich.: Philosophy of Science Association: 459–69.

————. 1995. *The Mangle of Practice*. Chicago: University of Chicago Press.

Piilonen, L., and A. Abashian. 1992. "On the Strength of the Evidence for the 17 keV Neutrino." *Progress in Atomic Physics, Neutrinos and Gravitation: Proceedings of the XXVIIth Rencontre de Moriond*. G. Chardin, O. Fackler, and J. Tran Thanh Van. Gif sur Yvette, France: Editions Frontieres: 225–34

Pinch, T. 1986. *Confronting Nature*. Dordrecht: Reidel.

Powell, C. F., P. F. Fowler, and D. H. Perkins. 1959. *The Study of Elementary Particles by the Photographic Method*. London: Pergamon Press.

Prescott, C. Y., W. B. Atwood, R. L. A. Cottrell, et al. 1978. "Parity Non-Conservation in Inelastic Electron Scattering." *Physics Letters B* 77: 347–52.

————. 1979. "Further Measurements of Parity Non-Conservation in Inelastic Electron Scattering." *Physics Letters B* 84: 524–28.

Raab, F. J. 1987. "Search for an Intermediate-Range Interaction: Results of the Eöt-Wash I Experiment." *New and Exotic Phenomena: Seventh Moriond Workshop*. O. Fackler and J. Tran Thanh Van. Gif sur Yvette, France: Editions Frontieres: 567–77.

Radcliffe, T., M. Chen, D. Imel, et al. 1992. "New Limits on the 17 keV Neutrino." *Progress in Atomic Physics, Neutrinos and Gravitation: Proceedings of the XXVIIth Rencontre de Moriond*. G. Chardin, O. Fackler, and J. Tran Thanh Van. Gif sur Yvette, France: Editions Frontieres: 217–24.

Randall, H. M., R. G. Fowler, N. Fuson, et al. 1949. *Infrared Determination of Organic Structures*. New York: Van Nostrand.

Rhein, M., B. Blank, E. Bozrk, et al. 1989. "Electron-Positron Coincidence Experiment at the TORI-Spectrometer." *Tests of Fundamental Laws in Physics*. O. Fackler and J. Tran Thanh Van. Gif sur Yvette: Editions Frontieres: 195–201.

Roe, B. P. 1992. *Probability and Statistics in Experimental Physics*. New York: Springer-Verlag.

Rutherford, E. 1913. *Radioactive Substances and their Radiations*. Cambridge: Cambridge University Press.

Sakai, M. 1989. "Reply to Comment on 'Electron Peaks in e^+ + Th, U, and Ta Interactions.'" *Physical Review C* 40: 1841.

Sakai, M., Y. Fujita, M. Imamura, et al. 1988. "Electron Peaks in e^+ + Th, U, and Ta Interactions." *Physical Review C* 38: 1971–74.

Salabura, P., H. Backe, K. Bethge, et al. 1990. "Correlated e^+e^- Peaks Observed in Heavy-ion Collisions." *Physics Letters B* 245: 153–60.

Schmidt, H. W. 1906. "Uber die Absorption der β-Strahlen des Radiums." *Physikalische Zeitschrift* 7: 764–66.

———. 1907. "Einige Versuche mit β-Strahlen von Radium E." *Physikalische Zeitschrift* 8: 361–73.

———. 1909. "Beitrag zur Frage uber den Durchgang der β -Strahlen durch Materie." *Physikalische Zeitschrift* 10: 929–48.

Schreckenbach, K. 1989. "Sensitive Search for Resonances in Bhabha Scattering around the Invariant Mass of 1.8 MeV/c^2." *Tests of Fundamental Laws in Physics.* O. Fackler and J. Tran Thanh Van. Gif sur Yvette: Editions Frontieres: 165–70.

Schwartz, A. J. 1995. "Why Do a Blind Analysis?" Princeton, N.J.: Princeton University (Internal report).

Schweppe, J., A. Gruppe, K. Bethge, et al. 1983. "Observation of a Peak Structure in Positron Spectra from U + Cm Collisions." *Physical Review Letters* 51: 2261–64.

Sellars, W. 1962. *Science, Perception, and Reality.* New York: Humanities Press.

Shapere, D. 1982. "The Concept of Observation in Science and Philosophy." *Philosophy of Science* 49: 482–525.

Shaviv, G., and J. Rosen, eds. 1975. *General Relativity and Gravitation: Proceedings of the Seventh International Conference (GR7), Tel-Aviv University, June 23–28, 1974.* New York: John Wiley.

Siegel, D. 1979. "Review of *The Scientific Imagination* by Gerald Holton." *American Journal of Physics* 47: 476–77.

Simpson, J. J. 1985. "Evidence of Heavy-Neutrino Emission in Beta Decay." *Physical Review Letters* 54: 1891–93.

———. 1986a. "Is There Evidence for a 17 keV Neutrino in the ^{35}S β Spectrum? The Case of Ohi et al." *Physics Letters B* 174: 113–14.

———. 1986b. "Evidence for a 17-keV Neutrino in ^{3}H and ^{35}S β Spectra." *'86 Massive Neutrinos in Astrophysics and in Particle Physics: Proceedings of the Sixth Moriond Workshop.* O. Fackler and J. Tran Thanh Van. Gif sur Yvette: Editions Frontieres: 565–77.

Simpson, J. J., and A. Hime. 1989. "Evidence of the 17-keV Neutrino in the β Spectrum of ^{35}S." *Physical Review D* 39: 1825–36.

Sona, P., M. Bini, T. Fazzini, et al. 1989. "A Sensitive Search for the Emission of a Neutral Particle in the Decay of the First Excited State in ^{16}O." *Tests of Fundamental Laws in Physics.* O. Fackler and J. Tran Thanh Van. Gif sur Yvette: Editions Frontieres: 217–20.

Speake, C. C., T. M. Niebauer, M. P. McHugh, et al. 1990. "Test of the Inverse-Square Law of Gravitation Using the 300-m Tower at Erie, Colorado." *Physical Review Letters* 65: 1967–71.

Stacey, F. D., G. J. Tuck, G. I. Moore, et al. 1987. "Geophysics and the Law of Gravity." *Reviews of Modern Physics* 59: 157–74.

Staley, K. 1999. "Golden Events and Statistics: What's Wrong with Galison's Image/Logic Distinction?" *Perspectives on Science* 7: 196–230.

Stubbs, C. W. 1990. "Seeking New Interactions: An Assessment and Overview." *New and Exotic Phenomena '90: Tenth Moriond Workshop.* O. Fackler and J. Tran Thanh Van. Gif sur Yvette, France: Editions Frontieres: 177–85.

Suppe, F. 1998a. "The Structure of a Scientific Paper." *Philosophy of Science* 65: 381–405.

————. 1998b. "Reply to Commentators." *Philosophy of Science* 65: 417–24.

Sur, B., E. B. Norman, K. T. Lesko, et al. 1991. "Evidence for the Emission of a 17-keV Neutrino in the β Decay of ^{14}C." *Physical Review Letters* 66: 2444–47.

Taubes, G. 1997. "The One That Got Away?" *Science* 275: 148–51.

Taylor, J. D., P.E.G. Baird, R. G. Hunt, et al. (1987). Parity Non-Conservation in Bismuth. Oxford University (Internal report).

Thieberger, P. 1987. "Search for a Substance-Dependent Force with a New Differential Accelerometer." *Physical Review Letters* 58: 1066–69.

————. 1989. "Thieberger Replies." *Physical Review Letters* 62: 810.

Thomas, J., P. Vogel, and P. Kasameyer. 1988. "Gravity Anomalies at the Nevada Test Site." *5th Force, Neutrino Physics: Eight Moriond Workshop.* O. Fackler and J. Tran Thanh Van. Gif sur Yvette, France: Editions Frontieres: 585–92.

Tsertos, H., E. Berdermann, F. Bosch, et al. 1985. "On the Scattering-Angle Dependence of the Monochromatic Positron Emission from U + U and U + Th Collisions." *Physics Letters B* 162: 273–76.

Tsertos, H., F. Bosch, P. Kienle, et al. 1987. "Multiple Positron Structures Observed from U-U Collisions at 5.6 MeV/u and 5.9 MeV/u." *Zeitschrift fur Physik A* 326: 235–36.

Tsertos, H., C. Kozhuharov, P. Armbruster, et al. 1988a. "Limits for Resonant Bhabha Scattering around the Invariant Mass of 1.8 MeV/c^2." *Zeitschrift fur Physik A* 331: 103–4.

————. 1988b. "Sensitive Search for Neutral Resonances in Bhabha Scattering around 1.8 MeV/c^2." *Physics Letters B* 207: 273–77.

Tuck, G. J. 1989. "Gravity Gradients at Mount Isa and Hilton Mines." *Abstracts of Contributed Papers, Twelfth International Conference on General Relativity and Gravitation*, July 2–8, 1989. Boulder, Colo.: University of Colorado.

van Fraassen, B. C. 1980. *The Scientific Image.* Oxford: Clarendon Press.

Van Klinken, J., W. J. Meiring, F. W. N. deBoer, et al. 1988. "Search for Resonant Bhabha Scattering Around an Invariant Mass of 1.8 MeV." *Physics Letters B* 205: 223–27.

Van Klinken, J., F. Bosch, U. Kneissl, et al. 1989. "Search for Resonant Bhabha Scattering." *Tests of Fundamental Laws in Physics.* O. Fackler and J. Tran Thanh Van. Gif sur Yvette: Editions Frontieres: 147–63.

von Baeyer, O., and O. Hahn. 1910. "Magnetic Line Spectra of Beta Rays." *Physikalische Zeitschrift* 11: 448–93.

von Baeyer, O., O. Hahn, and L. Meitner. 1911a. "Uber die β-Strahlen des aktiven Niederschlags des Thoriums." *Physikalische Zeitschrift* 12: 273–79.

————. 1911b. "Nachweis von β -Strahlen bei Radium D." *Physikalische Zeitschrift* 12: 378–79.

Von Wimmersperg, U., S. H. Connell, R. F. A. Hoernle, et al. 1987. "Observation of Bhabha Scattering in the Center-of-Mass Kinetic-Energy Range 342–845 keV." *Physical Review Letters* 59: 266–69.

Wang, T. F., I. Ahmad, S. J. Freedman, et al. 1987. "Search for Sharp Lines in e$^+$-e$^-$ Coincidences from Positrons on Th." *Physical Review C* 36: 2136–38.

Weber, J., M. Lee, D. J. Gretz, et al. 1973. "New Gravitational Radiation Experiments." *Physical Review Letters* 31: 779–83.

Wieman, C. E., S. L. Gilbert, and M. C. Noecker. 1987. "A New Measurement of Parity Nonconservation in Atomic Cesium." *Atomic Physics 10*. H. Narumi and I. Shimamura. Amsterdam: Elsevier Scientific: 65–76.

Wietfeldt, F. E., and E. B. Norman. 1996. "The 17 keV Neutrino." *Physics Reports* 273: 149–97.

Wietfeldt, F. E., Y. D. Chan, M. T. F. DaCruz, et al. 1993. "Further Studies of a ^{14}C-Doped Germanium Detector." *Bulletin of the American Physical Society* 38: 1855–56.

Wilson, W. 1909. "On the Absorption of Homogeneous β Rays by Matter, and on the Variation of the Absorption of the Rays with Velocity." *Proceedings of the Royal Society (London)* A82: 612–28.

———. 1910a. "The Decrease of Velocity of the β-Particles on Passing through Matter." *Proceedings of the Royal Society (London)* A84: 141–50.

———. 1910b. "Uber das Absorptionsgesetz der β-Strahlen." *Physikalische Zeitschrift* 11: 101–4.

———. 1912. "On the Absorption and Reflection of Homogneous β-Particles." *Proceedings of the Royal Society (London)* A87: 310–25.

Wiss, J., and R. Gardner. 1994. Estimating Systematic Errors. Urbana: University of Illinois.

Wu, X. Y., P. Asoka-Kumar, J. S. Greenberg, et al. 1992. "Search for Low-Mass States in e^+e^- Scattering." *Physical Review Letters* 69: 1729–32.

Zeitnitz, B., B. Armbruster, M. Becker, et al. 1998. "Neutrino Oscillation Results from Karmen." *Progress in Particle and Nuclear Physics* 40: 169–81.

Zlimen, I., S. Kaucic, and A. Ljubicic. 1988. "Search for Neutrinos with Masses in the Range of 16.4 → 17.4 keV." *Physica Scripta* 38: 539–42.

Zlimen, I., A. Ljubicic, S. Kaucic, et al. 1991. "Evidence for a 17-keV Neutrino." *Physical Review Letters* 67: 560–63.

Zweig, G. 1964. "An SU_3 Model for Strong Interaction Symmetry and Its Breaking." *CERN Reports* 8182/TH/401.

Index